PREFACE

This report was submitted by Committee 2 to the 1993–97 Commission in June 1994. The report was prepared jointly by the Task Group on Internal Dosimetry and the Task Group on Dose Calculations.

The Task Group on Internal Dosimetry had the following membership:

Members

J. W. Stather (Chairman)	R. A. Guilmette	H. Métivier
M. R. Bailey	J. D. Harrison	D. Noßke
A. Bouville	J. Inaba	M. Roy
F. T. Cross	R. W. Leggett	D. M. Taylor
K. F. Eckerman*		

Corresponding Members

J. C. Barton	F. A. Fry	J. Piechowski
P.-G. Beau	F. O. Hoffman	V. Repin
X. Chen	G. M. Kendall	M. Sikov
M. Cristy	N. Nelson	

The Task Group on Dose Calculations had the following membership:

Members

K. F. Eckerman (Chairman)	T. Dillman	R. W. Leggett
V. Berkovski	K. Henrichs	I. A. Likhtarev
M. Cristy	G. M. Kendall	D. Noßke

Corresponding Members

A. C. James	A. R. Reddy

During the period of preparation of this report, the membership of Committee 2 was:

A. Kaul (Chairman)	F. A. Fry	A. R. Reddy
A. Bouville	J. Inaba	M. Roy
X. Chen	I. A. Likhtarev	J. W. Stather
F. T. Cross	H. Métivier	D. M. Taylor
G. Dietze	H. G. Paretzke	R. H. Thomas
K. F. Eckerman		

Acknowledgements

The work of the Task Groups was greatly aided by significant technical contributions from A. Birchall, A. W. Phipps and T. P. Fell, National Radiological Protection Board, U.K.

* Chairman of the Dose Calculations and Reference Man Task Groups, is an '*ex officio*' member of the Task Group.

GLOSSARY OF TERMS

Absorbed Dose

the physical dose quantity given by

$$D = \frac{d\bar{\varepsilon}}{dm}$$

where $d\bar{\varepsilon}$ is the mean energy imparted by ionising radiation to the matter in a volume element and dm is the mass of the matter in this volume element. The SI unit for absorbed dose is joule per kilogram ($J\ kg^{-1}$) and its special name is gray (Gy).

Absorbed Fraction ($AF(T \leftarrow S)_R$)

the fraction of energy emitted as a specified radiation type R in a specified source region S which is absorbed in a specified target tissue T.

Becquerel (Bq)

the name for the SI unit of activity, $1\ Bq = 1\ s^{-1}$.

Cells Near Bone Surfaces

those tissues which lie within 10 μm of endosteal surfaces, and bone surfaces lined with epithelium.

Committed Effective Dose ($E(\tau)$)

the sum of the products of the committed organ or tissue equivalent doses and the appropriate organ or tissue weighting factors (w_T), where τ is the integration time in years following the intake. The integration time is 50 y for adults.

Committed Equivalent Dose ($H_T(\tau)$)

the time integral of the equivalent dose rate in a particular tissue or organ that will be received by an individual following intake of radioactive material into the body, where τ is the integration time in years following the intake. The integration time is 50 y for adults.

Cortical Bone

equivalent to "Compact Bone" in *ICRP Publication 20* (ICRP, 1973), that is, any bone with a surface/volume ratio less than $60\ cm^2\ cm^{-3}$; in Reference Man it has a mass of 4000 g.

Dose Coefficient

committed tissue equivalent dose per unit acute intake $h_T(\tau)$ or committed effective dose per unit acute intake $e(\tau)$, where τ is the time period in years over which the dose is calculated (e.g. $e(50)$).

Effective Dose (E)

the sum of the weighted equivalent doses in all tissues and organs of the body, given by the expression:

$$E = \sum_T w_T H_T$$

where H_T is the equivalent dose in tissue or organ, T, and w_T is the weighting factor for tissue T.

Equivalent Dose (H_T)
the equivalent dose, $H_{T,R}$, in tissue or organ T due to radiation R, is given by:

$$H_{T,R} = w_R D_{T,R}$$

where $D_{T,R}$ is the average absorbed dose from radiation R in tissue T and w_R is the radiation weighting factor. Since w_R is dimensionless, the units are the same as for absorbed dose, J kg^{-1}, and its special name is sievert (Sv). The total equivalent dose, H_T, is the sum of $H_{T,R}$ over all radiation types

$$H_T = \sum_R H_{T,R}$$

Fractional Absorption in the Gastrointestinal Tract (f_1)
the f_1 value is the fraction of an ingested element directly absorbed to body fluids.

Gray (Gy)
the special name for the SI unit of absorbed dose.
1 Gy = 1 J kg^{-1}.

Radiation Weighting Factor (w_R)
the radiation weighting factor is a dimensionless factor to derive the equivalent dose from the absorbed dose averaged over a tissue or organ and is based on the quality of radiation.

Red Bone Marrow (active)
the component of marrow which contains the bulk of the haematopoietic stem cells.

Reference Man
a person with the anatomical and physiological characteristics defined in the report of the ICRP Task Group on Reference Man (ICRP, 1975).

Sievert (Sv)
the name for the SI unit of equivalent dose and effective dose.
1 Sv = 1 J kg^{-1}.

Source Region (S)
region within the body containing the radionuclide. The region may be an organ, a tissue, the contents of the gastrointestinal tract or urinary bladder, or the surfaces of tissues as in the skeleton and the respiratory tract.

Specific Effective Energy ($SEE(T \leftarrow S)_R$)
the energy, suitably modified for radiation weighting factor, imparted per unit mass of a target tissue, T, as a consequence of the emission of a specified radiation, R, from a transformation occurring in source region S expressed as Sv $(Bq\ s)^{-1}$.

Target Tissue
tissue or organ in which radiation is absorbed.

Tissue Weighting Factor (w_T)
the factor by which the equivalent dose in a tissue or organ is weighted to represent the relative contributions of that tissue or organ to the total detriment resulting from uniform irradiation of the body.

Trabecular Bone
equivalent to "Cancellous Bone" in *ICRP Publication 20* (ICRP, 1973), i.e. any bone with a surface/volume ratio greater than 60 $cm^2\ cm^{-3}$; in Reference Man it has a mass of 1000 g.

Transfer Compartment
the compartment introduced (for mathematical convenience) into most of the biokinetic models used in this report to account for the translocation of the radioactive material through the body fluids from where they are deposited in tissues.

Respiratory Tract Model (ICRP, 1994)

Absorption
movement of material to blood regardless of mechanism. Generally applies to dissociation of particles and the uptake into blood of soluble substances and material dissociated from particles.

Aerodynamic diameter (d_{ae})
diameter (μm) of unit density sphere that has same terminal settling velocity in air as the particle of interest.

Alveolar-Interstitial Region (AI)
consists of the respiratory bronchioles, alveolar ducts and sacs with their alveoli, and the interstitial connective tissue; airway generations 16 and beyond.

AMAD
Activity Median Aerodynamic Diameter. Fifty percent of the activity in the aerosol is associated with particles of aerodynamic diameter (d_{ae}) greater than the AMAD. Used when deposition depends upon inertial impaction and sedimentation, typically when the AMAD is $>0.5\ \mu$m.

AMTD
Activity Median Thermodynamic Diameter. Fifty percent of the activity in the aerosol is associated with particles of thermodynamic diameter (d_{th}) greater than the AMTD. Used when deposition depends principally on diffusion, typically when AMAD is $<0.5\ \mu$m.

Basal cells
cuboidal epithelial cells attached to the basement membrane of extrathoracic and bronchial epithelium and not extending to the surface.

Bronchial Region (BB)
consists of the trachea (generation 0) and bronchi, generations 1 through 8.

Bronchiolar Region (bb)
consists of the bronchioles and terminal bronchioles; airway generations 9 through 15.

Clara Cells
nonciliated columnar epithelial cells in bronchioles that have serous secretions.

Class SR-1 Gases
soluble or reactive gases and vapours.

Class SR-2 Gases
highly soluble or reactive gases and vapours.

Class SR-0 Gases
insoluble and nonreactive gases and vapours.

Clearance
the removal of material from the respiratory tract by particle transport and by absorption into blood.

Compartments in the Respiratory Tract Model:

- LN_{ET} lymphatics and lymph nodes that drain the extrathoracic region.
- LN_{TH} lymphatics and lymph nodes that drain the thoracic region.
- BB_{bas} tissue in bronchial region through which basal cell nuclei are distributed.
- BB_{sec} tissue in bronchial region through which secretory cell nuclei are distributed.
- ET_{seq} compartment representing prolonged retention in airway tissue of small fraction of particles deposited in the nasal passages.
- BB_{seq} compartment representing prolonged retention in airway walls of small fraction of particles deposited in the bronchial region.
- bb_{seq} compartment representing prolonged retention in airway walls of small fraction of particles deposited in the bronchiolar region.

Deposition
refers to the initial processes determining how much of the material in the inspired air remains behind after expiration. Deposition of material may occur during both inspiration and expiration.

Extrathoracic (ET) Airways
consists of anterior nose (ET_1) and the posterior nasal passages, larynx, pharynx and mouth (ET_2).

Particle Transport
processes that clear material from the respiratory tract to the GI tract and to the lymph nodes, and move material from one part of the respiratory tract to another.

Secretory Cells
nonciliated epithelial cells that have mucous (mucus cells) or serous (Clara cells) secretions.

Thermodynamic diameter (d_{th})
diameter (μm) of a spherical particle that has the same diffusion coefficient in air as the particle of interest.

Thoracic (TH) Airways
combined Bronchial, Bronchiolar and Alveolar Interstitial regions.

Type F Materials
deposited materials that are readily absorbed into blood from the respiratory tract. (Fast rate of absorption.)

Type M Materials
deposited materials that have intermediate rates of absorption into blood from the respiratory tract. (Moderate rate of absorption.)

Type S Materials
deposited materials that are relatively insoluble in the respiratory tract. (Slow rate of absorption.)

References

ICRP (1973) *Alkaline Earth Metabolism in Adult Man*, ICRP Publication 20, Pergamon Press, Oxford.
ICRP (1975) *Report of the Task Group on Reference Man*, ICRP Publication 23, Pergamon Press, Oxford.
ICRP (1994) *Human Respiratory Tract Model for Radiological Protection*, ICRP Publication 66, *Annals of the ICRP* 24(1-3), Elsevier Science Ltd, Oxford.

1. INTRODUCTION

(1) The Commission's 1990 recommendations on radiation protection standards in *ICRP Publication 60* (ICRP, 1991a) were developed to take into account new biological information related to the detriment associated with radiation exposures and supersede the earlier recommendations in *ICRP Publication 26* (ICRP, 1977). Adoption of the recommendations necessitates a revision of the Commission's secondary limits contained in *ICRP Publication 30,* Parts 1–4 (ICRP, 1979a, 1980, 1981b, 1988b). In order to permit immediate application of these new recommendations, revised values of the Annual Limits on Intake (ALIs) based on the methodology and biokinetic information from *ICRP Publication 30,* but which incorporated the new dose limits and tissue weighting factors, w_T, were issued as *ICRP Publication 61* (ICRP, 1991b).

(2) Since issuing *ICRP Publication 61,* ICRP has published a revised kinetic and dosimetric model of the respiratory tract (ICRP, 1994). The main aim of the present report is to give values of dose coefficients for workers using this new model. The Commission has also issued, in *ICRP Publications 56, 67* and *69* (ICRP, 1989a, 1993a, 1995), new biokinetic models derived for selected radionuclides since the issue of *ICRP Publication 30.* Where revised biokinetic models and data have been given for adults these have been used in place of those given in *ICRP Publication 30* to compute the dose coefficients given in this report. The tissue and radiation weighting factors used in the calculations are those recommended in *ICRP Publication 60.* The present publication therefore replaces *ICRP Publication 61.* In due course a complete revision of *ICRP Publication 30* that takes into account new anatomical and physiological data, and newer biokinetic models will be issued.

2. RESPIRATORY TRACT MODEL

(3) The new Human Respiratory Tract Model for Radiological Protection (ICRP, 1994) constitutes an updating of the model used in *ICRP Publication 30* for workers. The model is described in full in *ICRP Publication 66* although a brief summary of the main features of the model is given here for adults. The new model takes into account extensive data on the behaviour of inhaled materials that have become available since the *ICRP Publication 30* model was developed (ICRP, 1966). As in the earlier model, deposition and clearance are treated separately (see below). The scope of the model has been extended to apply explicitly to all members of the population, giving reference values for 3-mo-old infants, 1-, 5-, 10- and 15-y-old children, and adults. The main difference in approach is that whereas the *ICRP Publication 30* model calculates only the average dose to the lungs, the new model calculates doses to specific tissues of the respiratory tract, and thus takes account of differences in radiosensitivity within the respiratory tract. In the new model, the respiratory tract is represented by five regions (Fig. 1). The extrathoracic (ET) airways are divided into ET_1, the anterior nasal passage and ET_2, which consists of the posterior nasal and oral passages, the pharynx and larynx. The thoracic regions are bronchial (BB: trachea, generation 0 and bronchi, airway generations 1–8), bronchiolar (bb: airway generations 9–15), and alveolar-interstitial

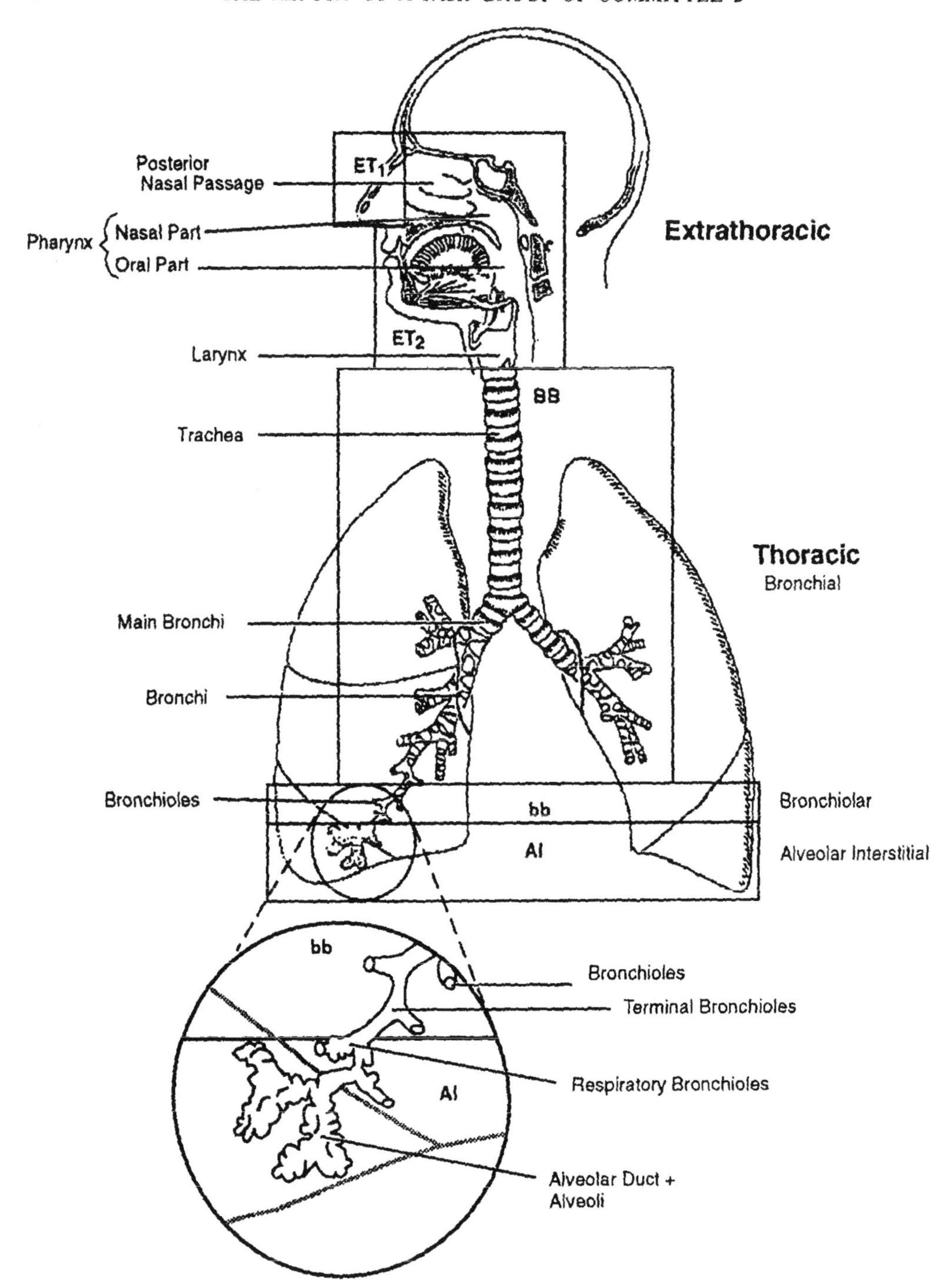

Fig. 1. Respiratory tract.

(AI: the gas exchange region). Lymphatic tissue is associated with the extrathoracic and thoracic airways (LN_{ET} and LN_{TH} respectively). Reference values of dimensions and scaling factors for subjects of different ages are specified.

2.1. Deposition

(4) The deposition model evaluates fractional deposition of an aerosol in each region, for all aerosol sizes of practical interest (0.6 nm–100 μm). For the ET regions, measured deposition efficiencies are related to characteristic parameters of particle size and airflow, and are scaled by anatomical dimensions to predict deposition under other conditions (e.g. age, gender, ethnic group). For the thoracic airways a theoretical model of gas transport and particle deposition is used to calculate particle deposition in each of the BB, bb and AI regions, and to quantify the effects of the subject's lung size and breathing rate. To model particle deposition, the regions are treated as a series of filters, during both inhalation and exhalation. The efficiency of each is evaluated by considering aerodynamic (gravitational settling, inertial impaction) and thermodynamic (diffusion) processes acting competitively. Regional deposition fractions are calculated for aerosols having log-normal particle size distributions, with geometric standard deviations (σ_g) taken to be a function of the median particle diameter, increasing from a value of 1.0 at 0.6 nm to a value of 2.5 above about 1 μm (*ICRP Publication 66*, Paragraph 170). Deposition parameters are given for four reference levels of activity (sleep, sitting, light exercise and heavy exercise).

(5) To calculate dose coefficients for inhalation of radionuclides by workers, the reference subject is taken to be a normal nose-breathing adult male at light work. For occupational exposure the default value now recommended for the Activity Median Aerodynamic Diameter (AMAD) is 5 μm (*ICRP Publication 66*, Paragraph 181), which is considered to be more representative of workplace aerosols than the 1 μm default value adopted in *ICRP Publication 30* (Dorrian and Bailey, 1995). Fractional deposition in each region of the respiratory tract of the reference worker is given in Table 1 for aerosols of 5 μm AMAD, together with values for 1 μm AMAD. In some situations, the smaller AMAD has been shown from field measurements to be more appropriate than the larger one.

2.2. Clearance

(6) The model describes several routes of clearance from the respiratory tract (Fig. 2). Material deposited in ET_1 is removed by extrinsic means such as nose-blowing. In other regions clearance is competitive between the movement of particles towards the gastrointestinal (GI) tract and lymph nodes (particle transport), and the absorption into blood of material from the particles in the respiratory tract. Clearance kinetics are expressed in terms of fractional clearance rates, i.e.:

$$\frac{dR_i(t)}{dt} = -\lambda_i(t)R_i(t) + m_{ji}(t)R_j(t)$$

where $R_i(t)$ is the amount of material retained in region i (ET_2, BB, bb or AI) at time t after intake, $\lambda_i(t)$ is the overall instantaneous rate of clearance of material from region i, and $m_{ji}(t)$ is the rate of clearance of material from any region j into region i.

Table 1. Regional deposition[a] of inhaled aerosols[b] in Reference Worker[c] (% of inhaled activity)

Region	1 μm	5 μm
ET_1	16.52	33.85
ET_2	21.12	39.91
BB[d]	1.24 (0.47055)	1.78 (0.33341)
bb[d]	1.65 (0.48926)	1.10 (0.39748)
AI	10.66	5.32
Total[e]	51.19	81.96

[a] Unless stated otherwise, the values in Tables 1–3, and in the following footnotes are reference values, i.e. the recommended values for use in the model. They are the exact values used to calculate the dose coefficients given later in this report and are therefore given to a greater degree of precision than would be chosen to reflect the certainty with which the average value of each parameter is known.

[b] The particles are assumed to have density 3.00 g cm^{-3}, and shape factor 1.5 (*ICRP Publication 66*, Paragraph 181). The aerosols are assumed to be log-normally distributed with geometric standard deviation, σ_g approximately 2.5. The value of σ_g is not a reference value, but is derived from the corresponding Activity Median Thermodynamic Diameter, AMTD (*ICRP Publication 66*, Paragraph 170).

[c] Light work is defined on the following basis: 2.5 h sitting (at which the amount inhaled is 0.54 m^3 h^{-1} and the breathing frequency 12 min^{-1}) and 5.5 h light exercise (at which the amount inhaled is 1.5 m^3 h^{-1} and the breathing frequency 20 min^{-1}) (*ICRP Publication 66*, Table 6). For both levels of activity all the inhaled air enters through the nose. The deposition fractions are therefore volume-weighted average values for the two levels of activity.

[d] It is assumed that a fraction of the deposits in BB and bb (0.007, independent of size) is retained in the airway wall (compartments BB_{seq} and bb_{seq} in Fig. 3). The fractions of the deposits in BB and bb that clear slowly (compartments BB_2 and bb_2 in Fig. 3) are given above in parentheses. These fractions are related to particle size and therefore depend on the size distributions of the particles deposited in the two regions (*ICRP Publication 66*, Paragraph 248).

[e] The total depositions of 1 and 5 μm aerosols in the lung model of *ICRP Publication 30* were about 63 and 91% respectively.

(7) It is assumed that the clearance rates due to particle transport and absorption to blood are independent. Thus the overall rate of clearance from a region is the sum of the rates due to the separate processes:

$$\begin{aligned}\lambda_i(t) &= m_i(t) + s_i(t)\\ &= g_i(t) + l_i(t) + s_i(t)\end{aligned}$$

where $m_i(t)$ and $s_i(t)$ are the clearance rates from region i due to particle transport and absorption, respectively; $g_i(t)$ and $l_i(t)$ are particle transport rates towards the GI tract and regional lymph nodes.

(8) The rates of clearance from each region, by each route, normally change with time after intake, and will in general be different for material deposited directly in the region during inhalation or cleared into a region following deposition in another region. Indeed, for the latter, the rate of particle transport out of region i depends on the time since the material was transported into the region, but its rate of absorption to blood depends on the time since the material originally deposited. Thus, $\lambda_i(t)$ and its components in the equations above are themselves dependent on the initial pattern of deposition, the time course of intake and the time course of transport from other regions into region i. To take account of this and to simplify calculations, clearance from each region is represented in the model by a combination of compartments. Each compartment clears at a constant fractional rate, such that the overall clearance approximates the required time-dependent behaviour.

(9) It is assumed that particle transport rates are the same for all materials. A single compartment model is therefore provided to describe particle transport of all materials (Fig. 3). Reference values of rate constants were derived, as far as possible, from human

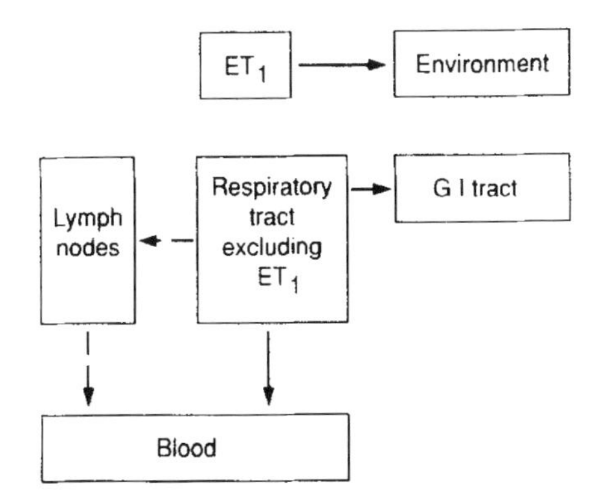

Fig. 2. Routes of clearance from the respiratory tract.

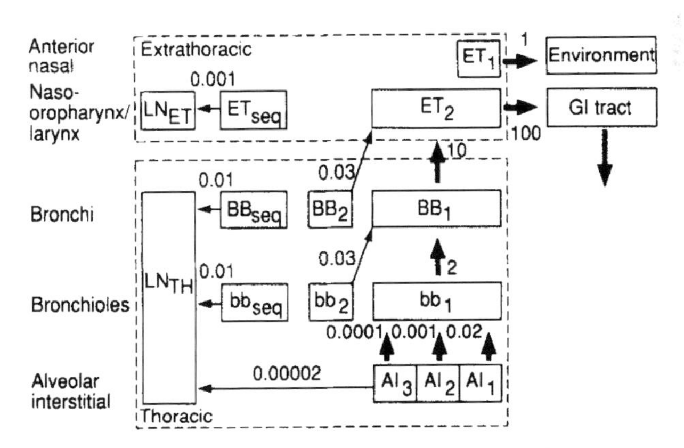

Fig. 3. Compartment model representing time-dependent particle transport from each respiratory tract region. Rates shown alongside arrows are in units of d^{-1}. It is assumed that (i) the AI deposit is divided between AI_1, AI_2 and AI_3 in the ratio 0.3:0.6:0.1; (ii) the fraction of the deposit in BB and bb that is cleared slowly (BB_2 and bb_2) is 50% for particles of physical diameter $< 2.5\ \mu m$ and decreases with diameter $> 2.5\ \mu m$, and the fraction retained in the airway wall (BB_{seq} and bb_{seq}) is 0.7% at all sizes; (iii) 0.05% of material deposited in ET_2 is retained in its wall (ET_{seq}). The model as shown above would describe the retention and clearance of a completely insoluble material. However, there is in general simultaneous absorption to blood of material from all the compartments except ET_1 (see text and Figs 4–6).

studies, since particle transport rates are known to vary greatly among mammalian species. Figure 3 as it stands would describe the retention and clearance of a completely insoluble material. However, as noted above, there is, in general, simultaneous absorption to blood.

(10) Absorption into blood depends on the physical and chemical form of the deposited material. It is assumed to occur at the same rate in all regions (including the lymph nodes) except ET_1, where it is assumed that none occurs. Absorption is a two-stage process: dissociation of the particles into material that can be absorbed into blood (dissolution); and absorption into blood of soluble material and of material dissociated from particles (uptake). The clearance rates associated with both stages can be time-dependent.

(11) The simplest compartment model representation of time-dependent *dissolution* is to assume that a fraction of the deposited material dissolves relatively rapidly, and the rest dissolves more slowly. The *ICRP Publication 66* model uses the system shown in Fig. 4 to represent this. In the model, the material deposited in the respiratory tract is assigned to compartments labelled "particles in initial state" in which it dissolves at a constant rate s_p. Material is simultaneously transferred (at a rate s_{pt}) to a corresponding compartment labelled "particles in transformed state" in which it has a different dissolution rate, s_t. (The ratio of s_p to s_{pt} approximates to the fraction that dissolves rapidly.) In different situations, the "particles in transformed state" may represent the residual material following dissolution of a relatively soluble component or surface layer, or material taken up by macrophages. The essential feature is that it remains subject to particle transport.

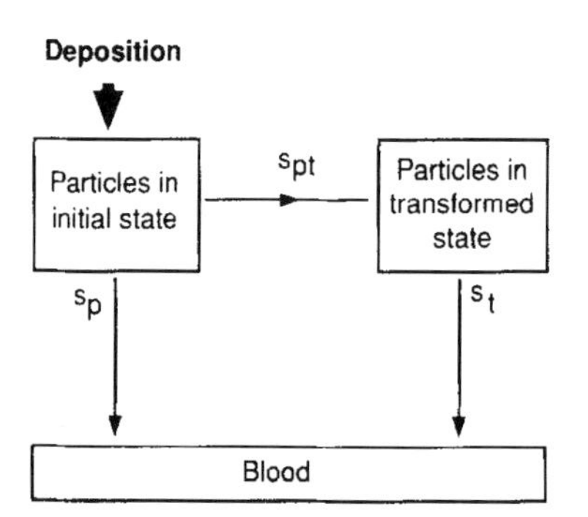

Fig. 4. Compartment model representing time-dependent dissolution followed by instantaneous uptake to blood. All the deposit is initially assigned to the compartment labelled "Particles in initial state". (For definition of symbols see text.)

(12) The system shown in Fig. 4 applies to each of the compartments in the particle transport compartment model shown in Fig. 3 (except ET_1). Thus from each of the 13 compartments containing "particles in initial state", material moves at a rate s_{pt} to a corresponding compartment k_T containing "particles in transformed state, T". The "particles in transformed state" are cleared by particle transport at the same rates as "particles in initial state". Thus if $m_{j,k}$ is the rate of particle transport from compartment j to compartment k containing "particles in initial state" and $m_{jT,kT}$ is the corresponding particle transport rate for "particles in transformed state", then $m_{j,k} = m_{jT,kT}$ for all j and k.

(13) Consider, for example, a compartment k, into which material moves by particle transport from compartment j at a rate $m_{j,k}$, and from which material moves by particle transport to compartment l at a rate $m_{k,l}$ (Fig. 5). The amount of material present in "particles in initial state", $I_k(t)$ at time t after intake, is described by:

$$\frac{dI_k(t)}{dt} = m_{j,k}I_j(t) - (m_{k,l} + s_p + s_{pt})I_k(t).$$

Similarly, the amount of material present in "particles in transformed state", $T_k(t)$ is described by:

$$\frac{dT_k(t)}{dt} = m_{j,k}T_j(t) + s_{pt}I_k(t) - (m_{k,l} + s_t)T_k(t).$$

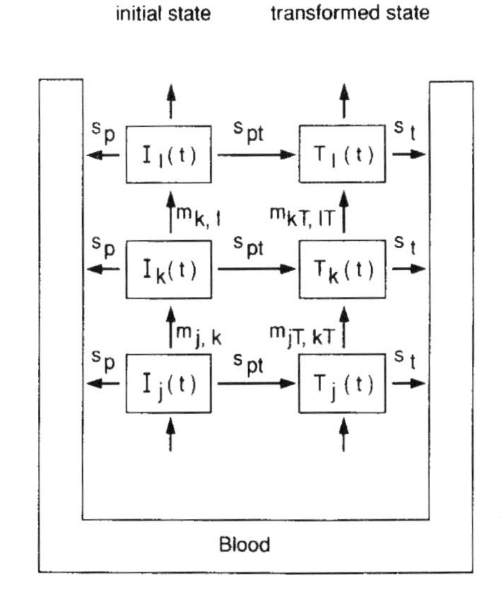

Fig. 5. Compartment model representing time-dependent particle transport and dissolution followed by instantaneous uptake to blood. Material moves into compartment k by particle transport from compartment j at a rate $m_{j,k}$, and from it by particle transport to compartment l at a rate $m_{k,l}$. $I_k(t)$ and $T_k(t)$ are the amounts of material present at time t in compartments representing "Particles in initial state" and "Particles in transformed state", respectively.

(14) For some compartments, such as bb_1 (Fig. 3), material enters by particle transport from more than one compartment and in the case of compartment AI_3, material is cleared by particle transport to two compartments. Additional appropriate components are required in the equations above to take account of this. In some cases, there is no compartment supplying material by particle transport (e.g. AI_1, AI_2, AI_3), and in others no particle transport out of the compartment (e.g. LN_{TH}, LN_{ET}) and correspondingly fewer components are required.

(15) *Uptake* to blood of dissociated material can usually be treated as instantaneous, but in some situations (as for certain gases and vapours, see below), a significant fraction of the dissociated material is absorbed slowly into blood as a result of binding to respiratory tract components. To represent time-dependent uptake, it is assumed that a fraction (f_b) of the dissolved material is retained in a "bound" state (Fig. 6), from

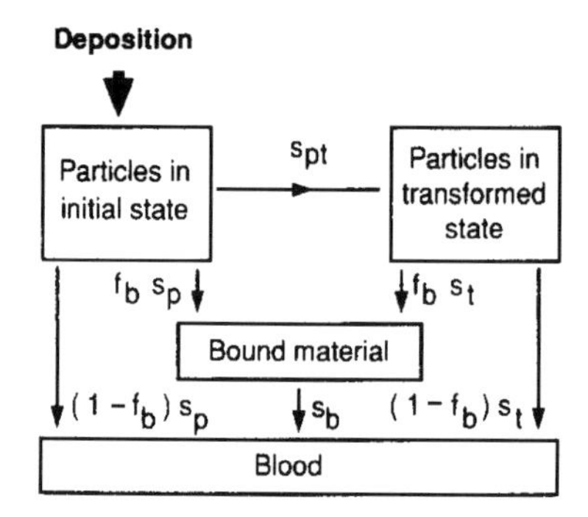

Fig. 6. Compartment model representing time-dependent absorption to blood (dissolution and uptake). Material in the compartment labelled "Particles in transformed state" is subject to particle transport at the same rate as material in the compartment labelled "Particles in initial state". Material in the compartment labelled "Bound material" is not subject to particle transport at all. It is cleared only by uptake into blood. (For definition of symbols see text.)

which it goes into blood at a rate s_b, while the remaining fraction $(1-f_b)$ goes to blood instantaneously. In the model, material in the "bound" state is not cleared by particle transport processes, but only by uptake to blood. Thus, only one "bound" compartment is required for each region.

(16) It is recommended that material-specific rates of absorption should be used in the respiratory tract model for compounds for which reliable human or animal experimental data exist. For other compounds, default parameters are recommended according to whether the absorption is considered to be fast (Type F), moderate (M) or slow (S) (corresponding broadly to inhalation Classes D, W and Y in the *ICRP Publication 30* system). Recommended values for each are specified in terms of the parameters s_p, s_{pt} and s_t, and are given in Table 2. The "bound" state is not invoked for the default parameters, i.e. $f_b = 0$ for all three types.

(17) Expressed as approximate half-times for one or two components of clearance, these absorption rates correspond to:

- Type F (fast): 10 min (100%);
- Type M (moderate): 10 min (10%); 140 d (90%);
- Type S (slow): 10 min (0.1%); 7000 d (99.9%).

(18) It is intended that in the replacement of *ICRP Publication 30* information relating to absorption in the respiratory tract of each element having isotopes of radiological importance will be reviewed. Where possible, absorption rates for important compounds will be recommended, and other compounds will be assigned to

Table 2. Default absorption rates for Type F, M, and S materials[a]

Type	F (fast)	M (moderate)	S (slow)
Model parameters (d^{-1}):			
s_p	100	10	0.1
s_{pt}	0	90	100
s_t	—	0.005	0.0001
f_b [b]	0	0	0
s_b [b]	—	—	—

[a] Reference values (see Footnote a to Table 1).
[b] No bound state assumed for default types.

the three absorption Types. For the purposes of this report, compounds for which clearance was given as Classes D, W or Y in the *ICRP Publication 30* system, are assigned to absorption Types F, M or S respectively (Annexe F).

2.3. Gases and Vapours

(19) For radionuclides inhaled in particulate form it is assumed that entry and deposition in the respiratory tract are governed by the size distribution of the aerosol particles. The situation is different for gases and vapours, for which the radionuclide has a specific behaviour at its site of entry to the respiratory tract, depending on the chemistry of the compound. Details of the treatment of gases and vapours are given in Annexe A.

(20) As for particulate material, the biokinetic behaviour of gases and vapours will be reviewed for the replacement of *ICRP Publication 30.* For this report the behaviour of gases and vapours assumed in *ICRP Publication 30* has been retained, but is represented using the formalism of *ICRP Publication 66.* The new model assigns gases and vapours to three classes (*ICRP Publication 66,* Paragraph 210):

- Class SR-1 (soluble or reactive). Consideration is given to retention in respiratory tract tissues, and to uptake to the systemic circulation, which may be less than 100% of the inhaled activity.
- Class SR-2 (highly soluble or reactive). For the examples considered here, there is complete and instantaneous systemic uptake of the inhaled activity.
- Class SR-0 (insoluble and non-reactive). Consideration is given to external irradiation from submersion in the cloud of gas, and to internal irradiation from gas within the respiratory tract.

2.4. Respiratory Tract Dosimetry

(21) The dose to the cells at risk in each region is given by the average dose to the target tissue in that region. The target cells identified in ET_1, ET_2, BB and bb, and the masses of tissue containing target cells in each region for dose calculations, are given in Table 3.

(22) In each of these regions there are also several possible sources. In bb, for example, activity in the fast phase of clearance (bb_1, Fig. 3) is taken to be in the mucous layer above the cilia; activity in the slow phase of clearance (bb_2) is taken to be in the fluid between the cilia; particles retained in the airway wall (bb_{seq}) are taken to be in a macrophage layer at a depth of 20–25 μm (i.e. below the target cells); activity "bound" to the epithelium is uniformly distributed in it; and account also has to be taken of irradiation from activity present in the AI region. For each source/target combination, *ICRP Publication 66* provides absorbed fractions for non-penetrating radiations: α, β and electrons; in each case as a function of energy. To obtain these, a single cylindrical geometry was used to represent each region of the conducting airways (ET_1, ET_2, BB, bb): the representative bronchus for BB being 5 mm diameter and the representative bronchiole for bb being 1 mm diameter (*ICRP Publication 66,* Paragraphs 48 and 54).

(23) To take account of differences in sensitivity between tissues, the equivalent dose, H_i, to each region, i, is multiplied by a factor, A_i, representing the region's sensitivity relative to that of the whole organ. The recommended values of A_i are also

Table 3. Target tissues of the respiratory tract

Organ tissue	Region	Target cells	Depth of target cell nuclei[a] (μm)	Mass of target tissue[a,b] (kg)	Assigned fraction[a,c] A_i of w_T
Extrathoracic airways	ET_1 (anterior nose)	Basal	40–50	2.0×10^{-5}	0.001
	ET_2 (posterior nose, larynx, pharynx, mouth)	Basal	40–50	4.5×10^{-4}	0.998
	LN_{ET} (lymphatics)		[d]	1.5×10^{-2}	0.001
Thoracic airways (lungs)	BB (bronchial)	Secretory (BB_{sec})	10–40	8.648×10^{-4}	0.333[c]
		Basal (BB_{bas})	35–50	4.324×10^{-4}	
	bb (bronchiolar)	Secretory	4–12	1.949×10^{-3}	0.333
	AI (alveolar–interstitial)		[d]	1.1	0.333
	LN_{TH} (lymphatics)		[d]	1.5×10^{-2}	0.001

[a] Reference values (see Footnote a to Table 1).
[b] Values for adult male. Masses for BB_{sec} and BB_{bas} are the masses of bronchial epithelium through which the nuclei of secretory cells and basal cells respectively are distributed and are based on reference values of airway dimensions. The mass of AI includes blood, but excludes lymph nodes.
[c] The dose to BB (H_{BB}) is calculated as the arithmetic mean of the doses to BB_{sec} and BB_{bas}.
[d] Average dose to region calculated.

given in Table 3. The weighted sum is the equivalent dose to the extrathoracic or thoracic airways respectively:

$$H_{ET} = H_{ET_1}A_{ET_1} + H_{ET_2}A_{ET_2} + H_{LN_{ET}}A_{LN_{ET}}$$

$$H_{TH} = H_{BB}A_{BB} + H_{bb}A_{bb} + H_{AI}A_{AI} + H_{LN_{TH}}A_{LN_{TH}}.$$

The tissue weighting factor, w_T of 0.12 specified for lung in *ICRP Publication 60*, is applied to the equivalent dose to the thoracic airways, H_{TH}. The extrathoracic airways are now included in the list of remainder tissues and organs (Section 6 and Table 9).

3. GASTROINTESTINAL TRACT MODEL

(24) The model used to describe the behaviour of radionuclides in the gastrointestinal (GI) tract and for the calculation of doses from radionuclides in the lumen of the gut is that described in *ICRP Publication 30*. In the 1990 recommendations of ICRP a tissue weighting factor is given for the colon which is not specifically identified in the GI tract model (see Table 7). In the calculations given in this report the dose to the colon is taken to be the mass weighted mean dose to upper and lower large intestine (ICRP, 1993a). The 1990 recommendations also include a tissue weighting factor for the oesophagus which is not included in the current ICRP model for the GI tract. In this report the specific absorbed fraction for the thymus is assumed to approximate that of the oesophagus (ICRP, 1991b, 1993).

(25) Since *ICRP Publication 30* further advice on the fractional uptake of elements from the GI tract (f_1 values) has been given by the ICRP in *Publications 56, 67* and *69* (ICRP, 1989a, 1993a, 1995) and by an Expert Group of the Nuclear Energy Agency of the OECD (NEA, 1988). The recommendations in these reports generally relate to the ingestion of radionuclides by members of the public and principally apply to radionuclides incorporated in foodstuffs, although in some cases better information on f_1 values would also be relevant to intakes in the workplace. The choice of appropriate f_1 values for radionuclides in the workplace is at present being reviewed by a Task Group of Committee 2 as part of the revision of *ICRP Publication 30*. The dose coefficients given in this report are based on the f_1 values given for occupational exposure in *ICRP Publication 30, Parts 1–4* (ICRP, 1979a, 1980, 1981b, 1988b) unless more appropriate values are given in *ICRP Publications 56, 67 and 69* (ICRP, 1989a, 1993a, 1995). Table 4 gives details of the f_1 values modified in this report from those given in *ICRP Publication 30*.

(26) Where material-specific information is available it should be used for calculating ingestion dose coefficients. In particular, radionuclides can in some circumstances be present in the workplace in insoluble matrices for which the f_1 values could be much lower than those given here.

(27) For the calculation of inhalation dose coefficients allowance has to be made for the absorption of material passing through the GI tract after clearance from the respiratory system. In this case, where f_1 values for ingestion have been changed from those given in *ICRP Publication 30* (Table 4), they are also adopted here for inhalation.

Table 4. f_1 values adopted for the calculation of dose coefficients[a]

Element	f_1	Chemical form	ICRP Publication
Co	0.05	Oxides and hydroxides	30
	0.1	All other compounds	67
Te	0.3	All compounds	67
Ce[b]	5×10^{-4}	All compounds	67
Th	2×10^{-4}	Oxides and hydroxides	30
	5×10^{-4}	All other compounds	69
U	2×10^{-3}	UO_2, U_3O_8 Most tetravalent compounds	30
	2×10^{-2}	All other compounds	69
Pu	1×10^{-4}	Nitrate	30
	1×10^{-5}	Oxide	30
	5×10^{-4}	All other compounds	67
Am, Np[c]	5×10^{-4}	All compounds	67

[a] For all other elements, *ICRP Publication 30* values are adopted.
[b] Also applied to all other lanthanides.
[c] Also applied to Ac, Pa, Cm, Bk, Cf, Es, Fm and Md.

4. BIOKINETIC MODELS

(28) The Commission has recently issued, in *ICRP Publications 56, 67* and *69* (ICRP, 1989a, 1993a, 1995), new biokinetic models that have been developed for selected radionuclides since the issue of *ICRP Publication 30.* The most substantial changes involve the alkaline earths and the actinides for which physiologically-based models have been developed. Where revised biokinetic models and data have been given for adults they have been used in place of those given in *ICRP Publication 30* to compute the dose coefficients given in this report and are specified in Table 5.

Table 5. Biokinetic models for systemic activity adopted for the calculation of dose coefficients[a]

Element	ICRP Publication	Element	ICRP Publication
Tritiated water	56	I	56
^3H as OBT[b]	56	Ba	67
Fe	69	Ce	67
Zn	67	Pb	67
Se	69	Po	67
Nb	56	Ra	67
Sr	67	Th	69
Zr	67	U	69
Mo	67	Np	67
Ag	67	Pu	67
Sb	69	Am	67
Te	67		

[a] For all other elements, *ICRP Publication 30* parameters are adopted.
[b] Organically bound tritium.

5. EXCRETION PATHWAYS FOR SYSTEMIC ACTIVITY

In the 1990 recommendations of ICRP, the urinary bladder and the colon are given explicit w_T values (see Table 7). Consequently, the equivalent doses from systemic activity being excreted in urine or faeces have been considered in this report.

(29) In *ICRP Publication 67* (ICRP, 1993a) recommended a biokinetic model for the urinary bladder based on the models of Snyder and Ford (1976) and Smith *et al.* (1982). In the *ICRP Publication 67* model, which is applied here, the adult bladder is taken to be of fixed size containing 115 ml of urine, which represents the average content of the bladder during the time period between voids. The volume of the bladder contents, noted above, corresponds to one-half the volume of urine excreted in a void (ICRP, 1975). To represent the kinetics of the bladder in terms of first-order processes, the rate of elimination from the bladder is taken to be twice the number of voids per day, taken to be 6. That is, the elimination rate from the bladder is taken to be 12 d^{-1} (equivalent to a half-time of about 1.4 h).

(30) The activity present in the upper and lower large intestine includes material which entered the GI tract from the systemic circulation. Except where otherwise indicated in biokinetic models, it is assumed for the calculation of doses that systemic activity lost in the faeces enters the GI tract via the upper large intestine.

(31) The rate of loss of systemic activity from the body through the routes of excretion is explicitly given in some of the biokinetic models of *ICRP Publications 67* and *69*, and it is expected that this practice will continue in the replacement of *ICRP Publication 30*. However, for this report it is necessary, as it was in some instances in *ICRP Publications 67* and *69*, to partition the systemic activity departing from the body between the routes of excretion according to a time-constant urinary to faecal excretion ratio. For the choice of these ratios the following priority ranking has been adopted from previous ICRP publications:

- *ICRP Publication 67* (ICRP, 1993a) and *69* (ICRP, 1995);
- *ICRP Publication 54* (ICRP, 1988a);
- *ICRP Publication 10/10A* (ICRP, 1968; 1971);
- *ICRP Publication 30* (ICRP, 1979a, 1980, 1981b, 1988b);
- where no guidance has been given in these publications a default urinary:faecal excretion ratio of 1 is applied.

(32) The sources of data on urinary to faecal excretion ratios used in the calculations given in this report are given in Table 6.

6. DOSE CALCULATIONS

(33) The Commission defines effective dose, E, as:

$$E = \sum_{T} w_T H_T$$

where w_T is the tissue weighting factor (Table 7) and H_T is the equivalent dose for tissue or organ T. The Commission now assigns explicit w_T values to a number of organs which, in *ICRP Publications 26* and *30*, were generally part of the remainder (colon, stomach,

Table 6. Excretion pathways adopted for systemic activity[e]

Element	Urinary to faecal excretion ratio: Ratio	ICRP Publication	Element	Urinary to faecal excretion ratio: Ratio	ICRP Publication
H	[a]	67	Tc	1	67
C	[a]	67	Ru	4	67
Na	100	10A	Ag	0.05	67
P	9	54	Sb	4	69
S	9	67	Te	4	67
Cl	100	10A	I	[b]	67
Ca	1	10	Cs	4	67
Fe	[b]	69	Ba	[b]	67
Co	6	67	Ce	0.11	67
Ni	20	67	Au	[d]	10
Zn	0.25	67	Pb	[b]	67
Ge	[c]	30	Po	0.5	67
Se	2	69	Ra	[b]	67
Rb	3	10	Th	[b]	69
Sr	[b]	67	U	[b]	69
Zr	5	67	Np	[b]	67
Nb	5	67	Pu	[b]	67
Mo	8	67	Am	[b]	67

[a] The excretion pathways are not considered in the recommended biokinetic parameters, bladder wall and upper and lower large intestine are assumed to receive the same dose as other tissues.

[b] The excretion pathways are considered explicitly by the recommended biokinetic model.

[c] Excretion only in urine for activity deposited in kidney. A ratio of 1 is assumed for activity in other tissues.

[d] Excretion only in urine is assumed.

[e] For other elements a ratio of 1 is assumed.

Table 7. Tissue weighting factors in *ICRP Publication 60*[a]

Organ or tissue (T)	Tissue weighting factor (w_T)
Gonads	0.20
Bone marrow (red)	0.12
Colon	0.12
Lung	0.12
Stomach	0.12
Bladder	0.05
Breast	0.05
Liver	0.05
Oesophagus	0.05
Thyroid	0.05
Skin	0.01
Bone surface	0.01
Remainder[b]	0.05

[a] The values have been developed from a reference population of equal numbers of both sexes and a wide range of ages. In the definition of effective dose they apply to workers, to the whole population, and to either sex.

[b] Now includes extrathoracic airways (see Table 9).

liver, oesophagus and urinary bladder). The committed effective dose over 50 y, $E(50)$, is obtained as the sum of the weighted committed equivalent doses to organs or tissues, $H_T(50)$.

(34) The committed (50 y) equivalent dose to organ or target tissue, T, is given by:

$$H_T(50) = \sum_S U_S(50)\ \mathrm{SEE}(T \leftarrow S).$$

$U_S(50)$ = Number of nuclear transformations (Bq s) in 50 y in source region S following an acute intake.

SEE(T←S) = Specific Effective Energy, equivalent dose in T per transformation in S expressed as (Sv(Bq s)$^{-1}$).

$$\mathrm{SEE}(T \leftarrow S) = \sum_R \frac{Y_R E_R w_R\ \mathrm{AF}(T \leftarrow S)_R}{m_T}.$$

Y_R = yield of radiation R per nuclear transformation (Bq s)$^{-1}$
E_R = energy of radiation R (J)
w_R = radiation weighting factor for radiation R (Table 8)
$\mathrm{AF}(T \leftarrow S)_R$ = absorbed fraction in T per transformation in S for radiation R
m_T = mass of target tissue, T (kg).

When a nuclide forms a decay product which itself is radioactive, the contribution of the decay product to the dose is evaluated using a similar set of equations for the decay product(s). The identity of decay products, information on the types of radiations emitted by the radionuclide (i.e. energies and intensities), and the physical half-lives of the radionuclides used in this report are those given in *ICRP Publication 38.*

(35) The usual assumption in *ICRP Publication 30* was that these radioactive decay products adopted the biokinetics of their parent; there were a few exceptions for decay products which were isotopes of iodine or noble gases. The new ICRP model of the respiratory tract proposes that the rate at which the particle dissociates is determined by its matrix, but the behaviour of dissociated material in the lung would be expected to depend on its elemental form. This consideration would also be relevant to decay products entering the systemic circulation. For the present implementation of the respiratory tract model, however, the absorption parameters of the parent are applied to all members of the decay chain. In this report separate systemic biokinetics have been applied to the parent and its decay products for those elements where this is assumed in *ICRP Publications 56, 67* and *69.* This applies to intakes of antimony, lead, radium,

Table 8. Radiation weighting factors in *ICRP Publication 60*

Type and energy range		Radiation weighting factor (w_R)
Photons, all energies		1
Electrons and muons, all energies		1
Neutrons, energy	< 10 keV	5
	> 10–100 keV	10
	> 100 keV–2MeV	20
	> 2–20 MeV	10
	> 20 MeV	5
Protons, energy	> 2 MeV	5
Alpha particles, fission fragments heavy nuclei		20

tellurium, thorium and uranium. For other elements, the treatment of decay products follows that in *ICRP Publication 30.*

(36)The dose coefficients presented in *ICRP Publication 30* were based on photon absorbed fractions, AF, published by Snyder *et al.* (1974) and which were also listed in the ICRP report on Reference Man, *ICRP Publication 23* (ICRP, 1975). More recent work at Oak Ridge National Laboratory has refined these calculations (Cristy and Eckerman, 1987, 1993) by including values for the female breast (for which muscle had previously been taken as a surrogate) and by more realistic modelling of bone and soft tissues. These updated absorbed fractions are used in all the calculations reported here. The assessment of doses from electrons and alpha particles is identical to that applied in *ICRP Publication 30* (except within the respiratory tract as described in *ICRP Publication 66*).

(37)The method of treating the remainder has been modified somewhat from that used in *ICRP Publication 30.* The remainder tissues and organs of *ICRP Publication 60* are given in Table 9. To this list has been added the extrathoracic airways as defined in the new model for the respiratory tract (Section 2).

Table 9. Masses of adult organs constituting the 10 remainder tissues at risk

Organ	Mass (g)
Muscle	28 000
Brain	1400
Small intestine	640
Kidneys	310
Pancreas	100
Spleen	180
Thymus	20
Uterus	80
Adrenals	14
Extrathoracic airways[a]	15
Total	30 759

[a] Mass taken from *ICRP Publication 66* (Table 5).

(38) The equivalent dose to remainder tissues, $H_{\text{remainder}}$, is normally calculated as the mass-weighted mean dose to the ten organs and tissues listed in Table 9. In the exceptional case in which the most highly irradiated remainder tissue or organ receives the highest equivalent dose of all organs, a weighting factor of 0.025 (half of remainder) is applied to that tissue or organ and 0.025 (half of remainder) to the mass-weighted equivalent dose in the rest of the remainder tissues and organs. The equivalent dose to remainder tissues is therefore calculated as:

$$H_{\text{remainder}} = \frac{\sum_{T=1}^{T=10} m_T H_T}{\sum_{T=1}^{T=10} m_T}, \qquad \text{if } H_{T'} \leq H_{\text{max}}$$

$$H_{\text{remainder}} = 0.5 \frac{\sum_{T=1(T \neq T')}^{T=10} m_T H_T}{\sum_{T=1(T \neq T')}^{T=10} m_T} + 0.5\, H_{T'}, \qquad \text{if } H_{T'} > H_{\text{max}}$$

where H_{max} is the highest equivalent dose to organs with explicit weighting factors, $H_{T'}$ is the equivalent dose of the remainder tissue or organ in which the equivalent dose is the highest of all remainder organs and m_T is the mass of that tissue, and where the summation extends over the ten remainder tissues.

(39) The committed effective dose $E(50)$ for the worker is then given by:

$$E(50) = \sum_{T=1}^{T=12} w_T H_T(50) + w_{\text{remainder}} H_{\text{remainder}}(50)$$

where $H_T(50)$ is the committed equivalent dose and w_T is the weighting factor for the 12 tissues and organs named in Table 7. $H_{remainder}(50)$ is the committed equivalent dose to the remainder tissues as defined above and $w_{remainder}$ is the tissue weighting factor (0.05) assigned to these remainder tissues.

(40) The dose coefficient $e(50)$ in Sv Bq^{-1} corresponds to the committed effective dose, $E(50)$, resulting from the intake of 1 Bq of that radionuclide.

7. SECONDARY LIMITS

(41) For occupational exposures, the 1990 recommendations of ICRP limit the effective dose to 100 mSv in a 5 y period (giving an average annual value of 20 mSv) with a limit of 50 mSv in any single year. The recommendations also include annual limits on the equivalent dose to skin and hands of 500 mSv and to the lens of the eye of 150 mSv. The latter are generally considered in relation to external radiation exposure of the body. In the case of internal exposure, the recommendations indicate that the ALI should be based on a committed effective dose of 20 mSv. The ALI (Bq) for any radionuclide can thus be obtained by dividing the annual average effective dose limit (0.02 Sv) by the dose coefficient $e(50)$.

$$\text{ALI} = \frac{0.02}{e(50)}$$

(42) Although ALIs based on the 1990 recommendations will be, for the most part, more restrictive than those based on the 1977 recommendations, an annual limitation on effective dose of 20 mSv permits intakes which can result in a committed equivalent dose of the order of 1 Sv to individual organs, in particular to bone surfaces, the kidneys and the extrathoracic tissues. Due to the protracted nature of the exposure from many internally deposited radionuclides, it is unlikely that the lifetime equivalent dose would be sufficient to result in deterministic effects. In addition, for the alpha-emitting radionuclides the committed effective dose (or the equivalent dose) as calculated here incorporates a radiation weighting factor, w_R, of 20 based on stochastic effects. As discussed in *ICRP Publication 58* (ICRP, 1989b), this probably overestimates the possibility of deterministic effects.

(43) For compliance with dose limits, the doses from external radiation as well as from ingested or inhaled radionuclides must be summed. The limitation on exposures may not necessarily be the ALI but the radiation dose from the cloud surrounding the individual.

8. TABLES OF DOSE COEFFICIENTS

(44) The dose coefficients for inhalation and ingestion of radionuclides using the biokinetic and dosimetric models described in this report are given in Annexes B, C and D. Annexes E and F give the f_1 values and lung clearance Types adopted for different chemical forms of the elements (see Sections 2.2 and 3).

(45) In *ICRP Publication 30*, it was pointed out that if the behaviour of any specific material were expected to differ significantly from that of the biokinetic model employed, then the model parameters should be modified to take account of the data available. The advice to use material-specific data is reinforced here, since the reference values chosen were selected to be representative rather than conservative. Where additional information is available, for example on particle size distribution or on absorption characteristics in the respiratory system, more accurate assessments can be made. The new respiratory tract model is designed to facilitate the application of model-specific parameter values for inhaled radionuclides.

(46) Dose coefficients for inhalation of particulates or for ingestion for about 800 radionuclides are given in Annexe B. These dose coefficients use the new respiratory tract model and either biokinetic models given in *ICRP Publication 30* or more recent models given in *ICRP Publications 56, 67* and *69* and detailed in Tables 4, 5 and 6. The effective dose is robust against changes in biokinetic and dosimetric models and the choices of parameters for individual organs usually have small effects on the effective dose per unit intake, generally within a factor of 2 or 3. Radon and its short-lived decay products are not covered here. They are dealt with separately in *ICRP Publication 65*.

(47) Annexe C gives dose coefficients for inhalation of soluble or reactive gases (Class SR-1 and SR-2). Class SR-2 gases and vapours are taken to be completely absorbed in the respiratory system and to distribute uniformly in body tissues, and for these the effective dose will tend to be the same as the "old" effective dose equivalent since both the old *(ICRP Publication 26)* and new *(ICRP Publication 60)* tissue weighting factors sum to one. There are differences in the mechanism of lung deposition under the new respiratory tract model and these are discussed above; however, transfer to blood is normally too rapid for there to be dosimetric implications.

(48) Annexe D gives effective dose rates for most gases which are insoluble and non-reactive (Class SR-0). These are effectively submersion doses. For noble gases, there are usually no significant changes from the doses given in *ICRP Publication 30* where the dominant contribution is external radiation from the semi-infinite cloud (submersion dose). However, the *ICRP Publication 60* scheme includes a weighting factor for the skin (0.01) and, for less penetrating radiation, this will tend to increase effective dose rate as compared to the old effective dose equivalent rate.

(49) The position is different if the radiation does not have sufficient energy to reach the basal layers of the skin. In these cases (e.g. ^{37}Ar) the submersion dose is negligible and the effective dose rate is dominated by the dose rate to the alveolar-interstitial (AI) region of the lung. Since the AI now has only one third of the total lung weighting factor,

the effective dose rate is about one third of the effective dose equivalent rate. The energy of the β particles for tritium gas are low enough to put it in the same category as ^{37}Ar. However, the dose from tritium gas is dominated by the small fraction which is taken to be metabolised. Tritium gas is, therefore, treated here as a Class SR-1 gas, and the dose coefficient is included in Annexe C.

References

Cristy, M. and Eckerman, K. F. (1987). *Specific absorbed fractions of energy at various ages from internal photon sources.* ORNL/TM-8381/V1-7. Oak Ridge National Laboratory, Oak Ridge, Tennessee, U.S.A.

Cristy, M. and Eckerman, K. F. (1993). *SEECAL: Program to calculate age-dependent specific effective energies.* ORNL/TM-12351. Oak Ridge National Laboratory, Oak Ridge, Tennessee, U.S.A.

Dorrian, M. D. and Bailey, M. R. (1995). *Particle size distributions of radioactive aerosols measured in workplaces. Radiat. Prot. Dosim.* (in press).

ICRP (1966). Task Group on Lung Dynamics. Deposition and retention models for internal dosimetry of the human respiratory tract. *Health Phys.* **12** (173–207).

ICRP (1968). *Evaluation of Radiation Doses to Body Tissues from Internal Contamination due to Occupational Exposure,* ICRP Publication 10. Pergamon Press, Oxford.

ICRP (1971). *The Assessment of Internal Contamination Resulting from Recurrent or Prolonged Uptakes,* ICRP Publication 10A. Pergamon Press, Oxford.

ICRP (1975). *Report of the Task Group on Reference Man,* ICRP Publication 23. Pergamon Press, Oxford.

ICRP (1977). *Recommendations of the ICRP,* ICRP Publication 26. *Annals of the ICRP* **1**(3). Pergamon Press, Oxford. Reprinted (with additions) in 1987.

ICRP (1979a). *Limits for Intakes of Radionuclides by Workers,* ICRP Publication 30, Part 1. *Annals of the ICRP* **2**(3/4). Pergamon Press, Oxford.

ICRP (1979b). *Limits for Intakes of Radionuclides by Workers,* ICRP Publication 30, Supplement to Part 1. *Annals of the ICRP* **3**(1–4). Pergamon Press, Oxford.

ICRP (1980). *Limits for Intakes of Radionuclides by Workers,* ICRP Publication 30, Part 2. *Annals of the ICRP* **4**(3/4). Pergamon Press, Oxford.

ICRP (1981a). *Limits for Intakes of Radionuclides by Workers,* ICRP Publication 30, Supplement to Part 2. *Annals of the ICRP* **5** (1–6). Pergamon Press, Oxford.

ICRP (1981b). *Limits for Intakes of Radionuclides by Workers,* ICRP Publication 30, Part 3 (including addendum to Parts 1 and 2). *Annals of the ICRP* **6**(2/3). Pergamon Press, Oxford.

ICRP (1982a). *Limits for Intakes of Radionuclides by Workers,* ICRP Publication 30, Supplement A to Part 3. *Annals of the ICRP* **7**(1–3). Pergamon Press, Oxford.

ICRP (1982b). *Limits for Intakes of Radionuclides by Workers,* ICRP Publication 30, Supplement B to Part 3 (including addendum to the Supplements of Parts 1 and 2). *Annals of the ICRP* **8**(1–3). Pergamon Press, Oxford.

ICRP (1983). *Radionuclide Transformations: Energy and Intensity of Emissions,* ICRP Publication 38. *Annals of the ICRP* **11–13**. Pergamon Press, Oxford.

ICRP (1986). *The Metabolism of Plutonium and Related Elements,* ICRP Publication 48. *Annals of the ICRP* **16**(2/3). Pergamon Press, Oxford.

ICRP (1988a). *Individual Monitoring for Intakes of Radionuclides by Workers: Design and Interpretation,* ICRP Publication 54. *Annals of the ICRP* **19**(1–3). Pergamon Press, Oxford.

ICRP(1988b). *Limits for Intakes of Radionuclides by Workers: An Addendum,* ICRP Publication 30, Part 4. *Annals of the ICRP* **19**(4). Pergamon Press, Oxford.

ICRP (1989a). *Age-Dependent Doses to Members of the Public from Intake of Radionuclides: Part 1,* ICRP Publication 56. *Annals of the ICRP* **20**(2). Pergamon Press, Oxford.

ICRP (1989b) *RBE for Deterministic Effects,* ICRP Publication 58. *Annals of the ICRP* **20**(4). Pergamon Press, Oxford.

ICRP (1991a). *1990 Recommendations of the International Commission on Radiological Protection,* ICRP Publication 60. *Annals of the ICRP* **21** (1–3). Pergamon Press, Oxford.

ICRP (1991b) *Annual Limits on Intake of Radionuclides by Workers Based on the 1990 Recommendations,* ICRP Publication 61. *Annals of the ICRP* **21**(4). Pergamon Press, Oxford.

ICRP (1993a) *Age-dependent Doses to Members of the Public from Intake of Radionuclides, Part 2: Ingestion Dose Coefficients,* ICRP Publication 67. *Annals of the ICRP* **23**(3/4). Elsevier Science Ltd., Oxford.

ICRP (1993b). *Protection Against Radon at Home and at Work,* ICRP Publication 65. *Annals of the ICRP* **23**(4). Elsevier Science Ltd., Oxford.

ICRP (1994). *Human Respiratory Tract Model for Radiological Protection,* ICRP Publication 66. *Annals of the ICRP* **24**(1–4). Elsevier Science Ltd., Oxford.

ICRP (1995). *Age-dependent Doses to Members of the Public from Intake of Radionuclides, Part 3: Ingestion Dose Coefficients*, ICRP Publication 69. Elsevier Science Ltd., Oxford. (To be published.)

NEA (OECD) (1988). Committee on Radiation Protection and Public Health, *Report of an Expert Group on Gut Transfer Factors*, NEA/OECD, Paris.

Smith, T., Veall, N. and Wootten, R. (1982). Bladder wall dose from administered radiopharmaceuticals: The effect on variation in urine flow rate, voiding interval and initial bladder content. *Radiat. Prot. Dosim.* **2** 183–189.

Snyder, W. S. and Ford, M. R. (1976). Estimation of doses to the urinary bladder and to the gonads. In: *Radiopharmaceutical Dosimetry Symposium* (Proc. Conf. Oak Ridge, Tennessee, April 1976). HEW Publication (FDA 76-8044), pp. 313–349. Department of Health, Education and Welfare, Bureau of Radiological Health, Rockville, MD.

ANNEXE A. TREATMENT OF GASES AND VAPOURS

(A1) Unlike deposition of particulate material, respiratory tract retention of gases and vapours is material-specific. Almost all inhaled gas molecules contact airway surfaces, but usually return to the air unless they dissolve in, or react with, the surface lining. The fraction of an inhaled gas or vapour that is retained in each region thus depends on its solubility and reactivity and, except in simple cases, has to be treated on an individual basis. As for particulate material, the biokinetic behaviour of gases and vapours will be reviewed in detail in the replacement of *ICRP Publication 30.* For this report the behaviour of gases and vapours assumed in *ICRP Publication 30* has been retained in most cases, but is represented using the formalism of *ICRP Publication 66.* Tritium in the form of hydrogen gas (HT) and iodine vapour are exceptions, see below. The new model assigns gases and vapours to three classes on the basis of the initial pattern of respiratory tract deposition (see the Glossary for definition of deposition).

- Class SR-1 (soluble or reactive: deposition throughout the respiratory tract, subsequent retention in the respiratory tract and absorption to blood determined by specific chemical properties of gas or vapour)
- Class SR-2 (highly soluble or reactive: deposition in ET)
- Class SR-0 (insoluble and non-reactive: negligible deposition)

Class SR-1

(A2) Soluble or reactive gases and vapours may irradiate all airways, and there may be significant exposure from absorption into tissues and blood. They require individual evaluation, and some guidance is given in *ICRP Publication 66.* It is recommended (*ICRP Publication 66,* Paragraph 221) that in the absence of information 100% deposition is assumed, with the following distribution: 10% ET_1, 20% ET_2, 10% BB, 20% bb, and 40% AI. For the purposes of this report, if total deposition differs from 100%, then regional deposits are by default distributed in the same proportions. It is also assumed that the distribution of the deposit in BB between the three compartments (Fig. 3) is the same as for particulate material of similar size, i.e. 0.007 to BB_{seq}, $f_s = 0.5$ to BB_2 and the balance $(1 - 0.007 - f_s)$ to BB_1. The same applies to the corresponding compartments in bb.

(A3) On the basis of the descriptions given in *ICRP Publication 30,* three compounds are assigned to Class SR-1: carbon monoxide, mercury vapour and nickel carbonyl. In each case, however, it is specified that all deposited material is subsequently transferred to blood. Since in the *ICRP Publication 66* model there is no absorption from ET_1, all the ET deposit is here assigned to ET_2.

Carbon monoxide

(A4) It is assumed in *ICRP Publication 30* that 40% of the inhaled activity is instantaneously bound to haemoglobin and the remaining 60% exhaled. The initial deposition is here taken to be 12% ET_2, 4% BB, 8% bb, 16% AI, but with instantaneous absorption to blood ($s_p = \infty$; $s_{pt} = 0$; $f_b = 0$; Fig. 6).

Mercury vapour

(A5) It is assumed in *ICRP Publication 30* that 70% of mercury entering the lung as mercury vapour is deposited there and subsequently translocated to blood with a half-time of 1.7 d. The initial deposition is therefore taken to be 10% BB, 20% bb and 40% AI. The deposited material is instantaneously transferred to the "bound" compartments in BB, bb and AI respectively ($s_p = \infty$; $s_{pt} = 0$; $f_b = 1$; Fig. 6). To give subsequent transfer to blood with a half-time of 1.7 d, s_b is set to $\ln 2/(1.7\ d) = 0.4\ d^{-1}$.

Nickel Carbonyl

(A6) It is assumed in *ICRP Publication 30* that all nickel entering the respiratory system is deposited in the lungs as nickel carbonyl and then translocated to the transfer compartment with a half-time of 0.1 d. The initial deposition is here taken to be 30% ET_2, 10% BB, 20% bb, 40% AI. The deposited material is instantaneously transferred to the "bound" compartments in ET, BB, bb and AI respectively ($s_p = \infty$; $s_{pt} = 0$; $f_b = 1$; Fig. 6). To give subsequent transfer to blood with a half-time of 0.1 d, s_b is set to $\ln 2/(0.1\ d) = 7\ d^{-1}$.

Elemental iodine

(A7) *ICRP Publication 30* does not address iodine vapour specifically, and therefore it was treated as a 1 μm AMAD Class D aerosol (ICRP, 1979a). Detailed studies of the deposition and subsequent biokinetics of inhaled elemental iodine conducted in human volunteers have indicated that nearly total deposition and subsequent absorption takes place (Black and Hounam, 1968; Morgan *et al.*, 1968). It was inferred that during normal nose-breathing there would be about 50% deposition in the extrathoracic airways and the remainder in the tracheo-bronchial region. The results suggested that there was some rapid absorption to blood directly from the respiratory tract, but also that much of the activity was swallowed and subsequently absorbed from the GI tract. On this basis, iodine vapour is assigned to Class SR-1, with deposition assumed to be 10% ET_1, 40% ET_2 and 50% BB, and with subsequent behaviour as Type F.

Class SR-2

(A8) Highly soluble or reactive gases and vapours are taken to be completely absorbed. In *ICRP Publication 30* there are a number of compounds for which it is assumed that all the inhaled material is completely and instantaneously translocated to "blood" or to the "transfer compartment" without a change in chemical form. These are:

- ^{3}H in organic compounds and in tritiated water. It should be noted that absorption through skin may need to be included in addition to the dose coefficients given in Annexe C.
- Sulphur as sulphur dioxide, carbonyl sulphide, hydrogen sulphide and carbon disulphide. Given as sulphur vapour in Annexe C.
- Carbon in all organic compounds and carbon dioxide. Given as carbon vapour in Annexe C.

(A9) Using the *ICRP Publication 66* model these compounds are classed as SR-2. For the purposes of calculation they are treated as if they were directly injected into the blood.

Class SR-0

(A10) Since for this class doses from absorbed gas are, by definition, negligible, internal doses are calculated assuming that the airways are uniformly filled with gas at the ambient concentration. Table A1 gives reference volumes of respiratory tract regions for this purpose and Annexe D gives effective dose rates (Sv d^{-1}/Bq m^{-3}) for Class SR-0 gases and vapours. They are based on the effective dose rates for submersion derived from Eckerman and Ryman (1993) and the lung equivalent dose rate from activity present in the lung volume modified by a w_T of 0.12. The treatment of decay products for the calculation of whole body and skin dose rates are as in *ICRP Publication 30.* No dose rates are calculated for decay products in lung volume.

Table A.1. Reference volumes of respiratory tract regions for calculating doses from gases within airways for Reference Worker[a]

Region	Volume (m^3)
ET_1	2.500×10^{-6}
ET_2	3.375×10^{-5}
BB	3.901×10^{-5}
bb	6.625×10^{-5}
AI	3.720×10^{-3}

[a] Values given to four significant figures for precision in calculation, see Footnote a to Table 1.

External irradiation

(A11) According to *ICRP Publication 30* (ICRP 1979, Paragraph 8.2.3) external radiation dominates exposure for all the radioisotopes of the noble gases Ar, Kr and Xe considered, except ^{37}Ar. This situation is not changed by application of the *ICRP Publication 66* respiratory tract model (Bailey *et al.,* 1995). However, the introduction of the w_T for skin of 0.01 in *ICRP Publication 60* (Table 5) will tend to result in an effective dose rate per unit exposure from external irradiation higher than the effective dose equivalent rate given in *ICRP Publication 30.*

Irradiation of the lung

(A12) *ICRP Publication 30* identifies ^{37}Ar and tritium in the form of hydrogen gas as two gases for which exposure is dominated by irradiation of the lung, because the emissions have insufficient energy to reach the basal layer of the skin. In both cases, the dose rate to the AI region is similar to the mean lung dose rate. However, since doses to BB and bb are lower, because of the short range of the emissions, the equivalent lung dose rate calculated with the *ICRP Publication 66* model is about one third of the mean lung dose rate calculated with the *ICRP Publication 30* model. Hence the effective dose rate is approximately one third of the effective dose equivalent rate. As noted in *ICRP Publication 66,* Paragraph 208, a small fraction of inhaled HT is absorbed and converted to HTO. On the assumption that 0.01% is so converted (Peterman *et al.,* 1985), the effective dose per unit intake from absorbed HT is 1.8×10^{-15} Sv Bq^{-1}, which is several times higher than that due to irradiation of the lung from gas within it

(Bailey, *et al.*, 1995). Tritium gas is therefore treated here as Class SR-1 and the dose coefficient is included in Annexe C.

References

Bailey, M. R., Birchall, A., Marsh, J. W., Phipps, A. and Sacoyannis, V. (1995). Application of the new ICRP Respiratory Tract Model to the treatment of gases and vapours in ICRP Publication 68. NRPB Memorandum (in preparation).

Black, A. and Hounam, R. F. (1968). Penetration of iodine vapour through the nose and mouth and the clearance and metabolism of the deposited iodine. *Ann. Occup. Hyg.* **11**, 209–225.

Eckerman, K. F. and Ryman, J. C. (1993). External Exposure to Radionuclides in Air, Water and Soil. Federal Guidance Report No. 12. Oak Ridge National Laboratory for U.S. Environmental Protection Agency. EPA 402-R-93-081.

ICRP (1979). *Limits for Intakes of Radionuclides by Workers.* ICRP Publication 30 Part 1. *Annals of the ICRP* **2**(3/4). Pergamon Press, Oxford.

ICRP (1991). *1990 Recommendations of the International Commission on Radiological Protection.* ICRP Publication 60. *Annals of the ICRP* **21** (1–3). Pergamon Press, Oxford.

Morgan, A., Morgan, D. J. and Black, A. (1968). A study of the deposition, translocation and excretion of radioiodine inhaled as iodine vapour. *Health Phys.* **15**, 313–322.

Peterman, B. F., Johnson, J. R. and McElroy, R. G. C. (1985). HT/HTO conversion in mammals. *Fusion Technol.* **8**, 2557–2563.

ANNEXE B. EFFECTIVE DOSE COEFFICIENTS FOR INGESTED AND INHALED PARTICULATES

Table B.1. Ingestion and inhalation of particulates

		Effective dose coefficients (Sv Bq^{-1})					
		Inhalation, $e_{inh}(50)$				Ingestion	
Nuclide	$t_{1/2}$	Type	f_1	1 μm AMAD	5 μm AMAD	f_1	$e_{ing}(50)$
Hydrogen							
Tritiated Water	12.3y	See Annex C for inhalation doses				1.000	1.8E-11
OBT	12.3y	See Annex C for inhalation doses				1.000	4.2E-11
Beryllium							
Be-7	53.3d	M	0.005	4.8E-11	4.3E-11	0.005	2.8E-11
		S	0.005	5.2E-11	4.6E-11		
Be-10	1.60E+06y	M	0.005	9.1E-09	6.7E-09	0.005	1.1E-09
		S	0.005	3.2E-08	1.9E-08		
Carbon							
C-11	0.340h	See Annex C for inhalation doses				1.000	2.4E-11
C-14	5.73E+03y	See Annex C for inhalation doses				1.000	5.8E-10
Fluorine							
F-18	1.83h	F	1.000	3.0E-11	5.4E-11	1.000	4.9E-11
		M	1.000	5.7E-11	8.9E-11		
		S	1.000	6.0E-11	9.3E-11		
Sodium							
Na-22	2.60y	F	1.000	1.3E-09	2.0E-09	1.000	3.2E-09
Na-24	15.0h	F	1.000	2.9E-10	5.3E-10	1.000	4.3E-10
Magnesium							
Mg-28	20.9h	F	0.500	6.4E-10	1.1E-09	0.500	2.2E-09
		M	0.500	1.2E-09	1.7E-09		
Aluminium							
Al-26	7.16E+05y	F	0.010	1.1E-08	1.4E-08	0.010	3.5E-09
		M	0.010	1.8E-08	1.2E-08		
Silicon							
Si-31	2.62h	F	0.010	2.9E-11	5.1E-11	0.010	1.6E-10
		M	0.010	7.5E-11	1.1E-10		
		S	0.010	8.0E-11	1.1E-10		

OBT—Organically bound tritium.

Table B.1.—(*continued*)

		Effective dose coefficients (Sv Bq^{-1})					
		Inhalation, $e_{inh}(50)$				Ingestion	
Nuclide	$t_{1/2}$	Type	f_1	1 μm AMAD	5 μm AMAD	f_1	$e_{ing}(50)$
Si-32	4.50E+02y	F	0.010	3.2E-09	3.7E-09	0.010	5.6E-10
		M	0.010	1.5E-08	9.6E-09		
		S	0.010	1.1E-07	5.5E-08		
Phosphorus							
P-32	14.3d	F	0.800	8.0E-10	1.1E-09	0.800	2.4E-09
		M	0.800	3.2E-09	2.9E-09		
P-33	25.4d	F	0.800	9.6E-11	1.4E-10	0.800	2.4E-10
		M	0.800	1.4E-09	1.3E-09		
Sulphur							
S-35 (inorganic)	87.4d	F	0.800	5.3E-11	8.0E-11	0.800	1.4E-10
		M	0.800	1.3E-09	1.1E-09	0.100	1.9E-10
S-35 (organic)	87.4d	See Annex C for inhalation doses				1.000	7.7E-10
Chlorine							
Cl-36	3.01E+05y	F	1.000	3.4E-10	4.9E-10	1.000	9.3E-10
		M	1.000	6.9E-09	5.1E-09		
Cl-38	0.620h	F	1.000	2.7E-11	4.6E-11	1.000	1.2E-10
		M	1.000	4.7E-11	7.3E-11		
Cl-39	0.927h	F	1.000	2.7E-11	4.8E-11	1.000	8.5E-11
		M	1.000	4.8E-11	7.6E-11		
Potassium							
K-40	1.28E+09y	F	1.000	2.1E-09	3.0E-09	1.000	6.2E-09
K-42	12.4h	F	1.000	1.3E-10	2.0E-10	1.000	4.3E-10
K-43	22.6h	F	1.000	1.5E-10	2.6E-10	1.000	2.5E-10
K-44	0.369h	F	1.000	2.1E-11	3.7E-11	1.000	8.4E-11
K-45	0.333h	F	1.000	1.6E-11	2.8E-11	1.000	5.4E-11
Calcium							
Ca-41	1.40E+05y	M	0.300	1.7E-10	1.9E-10	0.300	2.9E-10
Ca-45	163d	M	0.300	2.7E-09	2.3E-09	0.300	7.6E-10

Table B.1.—(*continued*)

		Effective dose coefficients (Sv Bq^{-1})					
		Inhalation, $e_{inh}(50)$				Ingestion	
Nuclide	$t_{1/2}$	Type	f_1	1 μm AMAD	5 μm AMAD	f_1	$e_{ing}(50)$
Ca-47	4.53d	M	0.300	1.8E-09	2.1E-09	0.300	1.6E-09
Scandium							
Sc-43	3.89h	S	1.0E-04	1.2E-10	1.8E-10	1.0E-04	1.9E-10
Sc-44	3.93h	S	1.0E-04	1.9E-10	3.0E-10	1.0E-04	3.5E-10
Sc-44m	2.44d	S	1.0E-04	1.5E-09	2.0E-09	1.0E-04	2.4E-09
Sc-46	83.8d	S	1.0E-04	6.4E-09	4.8E-09	1.0E-04	1.5E-09
Sc-47	3.35d	S	1.0E-04	7.0E-10	7.3E-10	1.0E-04	5.4E-10
Sc-48	1.82d	S	1.0E-04	1.1E-09	1.6E-09	1.0E-04	1.7E-09
Sc-49	0.956h	S	1.0E-04	4.1E-11	6.1E-11	1.0E-04	8.2E-11
Titanium							
Ti-44	47.3y	F	0.010	6.1E-08	7.2E-08	0.010	5.8E-09
		M	0.010	4.0E-08	2.7E-08		
		S	0.010	1.2E-07	6.2E-08		
Ti-45	3.08h	F	0.010	4.6E-11	8.3E-11	0.010	1.5E-10
		M	0.010	9.1E-11	1.4E-10		
		S	0.010	9.6E-11	1.5E-10		
Vanadium							
V-47	0.543h	F	0.010	1.9E-11	3.2E-11	0.010	6.3E-11
		M	0.010	3.1E-11	5.0E-11		
V-48	16.2d	F	0.010	1.1E-09	1.7E-09	0.010	2.0E-09
		M	0.010	2.3E-09	2.7E-09		
V-49	330d	F	0.010	2.1E-11	2.6E-11	0.010	1.8E-11
		M	0.010	3.2E-11	2.3E-11		
Chromium							
Cr-48	23.0h	F	0.100	1.0E-10	1.7E-10	0.100	2.0E-10
		M	0.100	2.0E-10	2.3E-10	0.010	2.0E-10
		S	0.100	2.2E-10	2.5E-10		
Cr-49	0.702h	F	0.100	2.0E-11	3.5E-11	0.100	6.1E-11
		M	0.100	3.5E-11	5.6E-11	0.010	6.1E-11
		S	0.100	3.7E-11	5.9E-11		

Table B.1.—(*continued*)

		Effective dose coefficients (Sv Bq^{-1})					
		Inhalation, $e_{inh}(50)$				Ingestion	
Nuclide	$t_{1/2}$	Type	f_1	1 μm AMAD	5 μm AMAD	f_1	$e_{ing}(50)$
Cr-51	27.7d	F	0.100	2.1E-11	3.0E-11	0.100	3.8E-11
		M	0.100	3.1E-11	3.4E-11	0.010	3.7E-11
		S	0.100	3.6E-11	3.6E-11		
Manganese							
Mn-51	0.770h	F	0.100	2.4E-11	4.2E-11	0.100	9.3E-11
		M	0.100	4.3E-11	6.8E-11		
Mn-52	5.59d	F	0.100	9.9E-10	1.6E-09	0.100	1.8E-09
		M	0.100	1.4E-09	1.8E-09		
Mn-52m	0.352h	F	0.100	2.0E-11	3.5E-11	0.100	6.9E-11
		M	0.100	3.0E-11	5.0E-11		
Mn-53	3.70E+06y	F	0.100	2.9E-11	3.6E-11	0.100	3.0E-11
		M	0.100	5.2E-11	3.6E-11		
Mn-54	312d	F	0.100	8.7E-10	1.1E-09	0.100	7.1E-10
		M	0.100	1.5E-09	1.2E-09		
Mn-56	2.58h	F	0.100	6.9E-11	1.2E-10	0.100	2.5E-10
		M	0.100	1.3E-10	2.0E-10		
Iron							
Fe-52	8.28h	F	0.100	4.1E-10	6.9E-10	0.100	1.4E-09
		M	0.100	6.3E-10	9.5E-10		
Fe-55	2.70y	F	0.100	7.7E-10	9.2E-10	0.100	3.3E-10
		M	0.100	3.7E-10	3.3E-10		
Fe-59	44.5d	F	0.100	2.2E-09	3.0E-09	0.100	1.8E-09
		M	0.100	3.5E-09	3.2E-09		
Fe-60	1.00E+05y	F	0.100	2.8E-07	3.3E-07	0.100	1.1E-07
		M	0.100	1.3E-07	1.2E-07		
Cobalt							
Co-55	17.5h	M	0.100	5.1E-10	7.8E-10	0.100	1.0E-09
		S	0.050	5.5E-10	8.3E-10	0.050	1.1E-09
Co-56	78.7d	M	0.100	4.6E-09	4.0E-09	0.100	2.5E-09
		S	0.050	6.3E-09	4.9E-09	0.050	2.3E-09
Co-57	271d	M	0.100	5.2E-10	3.9E-10	0.100	2.1E-10
		S	0.050	9.4E-10	6.0E-10	0.050	1.9E-10

Table B.1.—(*continued*)

		Effective dose coefficients (Sv Bq^{-1})					
		Inhalation, $e_{inh}(50)$				Ingestion	
Nuclide	$t_{1/2}$	Type	f_1	1 μm AMAD	5 μm AMAD	f_1	$e_{ing}(50)$
Co-58	70.8d	M	0.100	1.5E-09	1.4E-09	0.100	7.4E-10
		S	0.050	2.0E-09	1.7E-09	0.050	7.0E-10
Co-58m	9.15h	M	0.100	1.3E-11	1.5E-11	0.100	2.4E-11
		S	0.050	1.6E-11	1.7E-11	0.050	2.4E-11
Co-60	5.27y	M	0.100	9.6E-09	7.1E-09	0.100	3.4E-09
		S	0.050	2.9E-08	1.7E-08	0.050	2.5E-09
Co-60m	0.174h	M	0.100	1.1E-12	1.2E-12	0.100	1.7E-12
		S	0.050	1.3E-12	1.2E-12	0.050	1.7E-12
Co-61	1.65h	M	0.100	4.8E-11	7.1E-11	0.100	7.4E-11
		S	0.050	5.1E-11	7.5E-11	0.050	7.4E-11
Co-62m	0.232h	M	0.100	2.1E-11	3.6E-11	0.100	4.7E-11
		S	0.050	2.2E-11	3.7E-11	0.050	4.7E-11
Nickel							
Ni-56	6.10d	F	0.050	5.1E-10	7.9E-10	0.050	8.6E-10
		M	0.050	8.6E-10	9.6E-10		
Ni-57	1.50d	F	0.050	2.8E-10	5.0E-10	0.050	8.7E-10
		M	0.050	5.1E-10	7.6E-10		
Ni-59	7.50E+04y	F	0.050	1.8E-10	2.2E-10	0.050	6.3E-11
		M	0.050	1.3E-10	9.4E-11		
Ni-63	96.0y	F	0.050	4.4E-10	5.2E-10	0.050	1.5E-10
		M	0.050	4.4E-10	3.1E-10		
Ni-65	2.52h	F	0.050	4.4E-11	7.5E-11	0.050	1.8E-10
		M	0.050	8.7E-11	1.3E-10		
Ni-66	2.27d	F	0.050	4.5E-10	7.6E-10	0.050	3.0E-09
		M	0.050	1.6E-09	1.9E-09		
Copper							
Cu-60	0.387h	F	0.500	2.4E-11	4.4E-11	0.500	7.0E-11
		M	0.500	3.5E-11	6.0E-11		
		S	0.500	3.6E-11	6.2E-11		
Cu-61	3.41h	F	0.500	4.0E-11	7.3E-11	0.500	1.2E-10
		M	0.500	7.6E-11	1.2E-10		
		S	0.500	8.0E-11	1.2E-10		

Table B.1.—(*continued*)

Nuclide	$t_{1/2}$	Effective dose coefficients (Sv Bq^{-1})				Ingestion	
		Inhalation, $e_{inh}(50)$					
		Type	f_1	1 μmAMAD	5 μmAMAD	f_1	$e_{ing}(50)$
Cu-64	12.7h	F	0.500	3.8E-11	6.8E-11	0.500	1.2E-10
		M	0.500	1.1E-10	1.5E-10		
		S	0.500	1.2E-10	1.5E-10		
Cu-67	2.58d	F	0.500	1.1E-10	1.8E-10	0.500	3.4E-10
		M	0.500	5.2E-10	5.3E-10		
		S	0.500	5.8E-10	5.8E-10		
Zinc							
Zn-62	9.26h	S	0.500	4.7E-10	6.6E-10	0.500	9.4E-10
Zn-63	0.635h	S	0.500	3.8E-11	6.1E-11	0.500	7.9E-11
Zn-65	244d	S	0.500	2.9E-09	2.8E-09	0.500	3.9E-09
Zn-69	0.950h	S	0.500	2.8E-11	4.3E-11	0.500	3.1E-11
Zn-69m	13.8h	S	0.500	2.6E-10	3.3E-10	0.500	3.3E-10
Zn-71m	3.92h	S	0.500	1.6E-10	2.4E-10	0.500	2.4E-10
Zn-72	1.94d	S	0.500	1.2E-09	1.5E-09	0.500	1.4E-09
Gallium							
Ga-65	0.253h	F	0.001	1.2E-11	2.0E-11	0.001	3.7E-11
		M	0.001	1.8E-11	2.9E-11		
Ga-66	9.40h	F	0.001	2.7E-10	4.7E-10	0.001	1.2E-09
		M	0.001	4.6E-10	7.1E-10		
Ga-67	3.26d	F	0.001	6.8E-11	1.1E-10	0.001	1.9E-10
		M	0.001	2.3E-10	2.8E-10		
Ga-68	1.13h	F	0.001	2.8E-11	4.9E-11	0.001	1.0E-10
		M	0.001	5.1E-11	8.1E-11		
Ga-70	0.353h	F	0.001	9.3E-12	1.6E-11	0.001	3.1E-11
		M	0.001	1.6E-11	2.6E-11		
Ga-72	14.1h	F	0.001	3.1E-10	5.6E-10	0.001	1.1E-09
		M	0.001	5.5E-10	8.4E-10		
Ga-73	4.91h	F	0.001	5.8E-11	1.0E-10	0.001	2.6E-10
		M	0.001	1.5E-10	2.0E-10		

Table B.1.—(*continued*)

		Effective dose coefficients (Sv Bq^{-1})					
		Inhalation, $e_{inh}(50)$				Ingestion	
Nuclide	$t_{1/2}$	Type	f_1	1 μm AMAD	5 μm AMAD	f_1	$e_{ing}(50)$
Germanium							
Ge-66	2.27h	F	1.000	5.7E-11	9.9E-11	1.000	1.0E-10
		M	1.000	9.2E-11	1.3E-10		
Ge-67	0.312h	F	1.000	1.6E-11	2.8E-11	1.000	6.5E-11
		M	1.000	2.6E-11	4.2E-11		
Ge-68	288d	F	1.000	5.4E-10	8.3E-10	1.000	1.3E-09
		M	1.000	1.3E-08	7.9E-09		
Ge-69	1.63d	F	1.000	1.4E-10	2.5E-10	1.000	2.4E-10
		M	1.000	2.9E-10	3.7E-10		
Ge-71	11.8d	F	1.000	5.0E-12	7.8E-12	1.000	1.2E-11
		M	1.000	1.0E-11	1.1E-11		
Ge-75	1.38h	F	1.000	1.6E-11	2.7E-11	1.000	4.6E-11
		M	1.000	3.7E-11	5.4E-11		
Ge-77	11.3h	F	1.000	1.5E-10	2.5E-10	1.000	3.3E-10
		M	1.000	3.6E-10	4.5E-10		
Ge-78	1.45h	F	1.000	4.8E-11	8.1E-11	1.000	1.2E-10
		M	1.000	9.7E-11	1.4E-10		
Arsenic							
As-69	0.253h	M	0.500	2.2E-11	3.5E-11	0.500	5.7E-11
As-70	0.876h	M	0.500	7.2E-11	1.2E-10	0.500	1.3E-10
As-71	2.70d	M	0.500	4.0E-10	5.0E-10	0.500	4.6E-10
As-72	1.08d	M	0.500	9.2E-10	1.3E-09	0.500	1.8E-09
As-73	80.3d	M	0.500	9.3E-10	6.5E-10	0.500	2.6E-10
As-74	17.8d	M	0.500	2.1E-09	1.8E-09	0.500	1.3E-09
As-76	1.10d	M	0.500	7.4E-10	9.2E-10	0.500	1.6E-09
As-77	1.62d	M	0.500	3.8E-10	4.2E-10	0.500	4.0E-10
As-78	1.51h	M	0.500	9.2E-11	1.4E-10	0.500	2.1E-10

Table B.1.—(*continued*)

		Effective dose coefficients (Sv Bq^{-1})					
		Inhalation, $e_{inh}(50)$				Ingestion	
Nuclide	$t_{1/2}$	Type	f_1	1 μm AMAD	5 μm AMAD	f_1	$e_{ing}(50)$
Selenium							
Se-70	0.683h	F	0.800	4.5E-11	8.2E-11	0.800	1.2E-10
		M	0.800	7.3E-11	1.2E-10	0.050	1.4E-10
Se-73	7.15h	F	0.800	8.6E-11	1.5E-10	0.800	2.1E-10
		M	0.800	1.6E-10	2.4E-10	0.050	3.9E-10
Se-73m	0.650h	F	0.800	9.9E-12	1.7E-11	0.800	2.8E-11
		M	0.800	1.8E-11	2.7E-11	0.050	4.1E-11
Se-75	120d	F	0.800	1.0E-09	1.4E-09	0.800	2.6E-09
		M	0.800	1.4E-09	1.7E-09	0.050	4.1E-10
Se-79	6.50E+04y	F	0.800	1.2E-09	1.6E-09	0.800	2.9E-09
		M	0.800	2.9E-09	3.1E-09	0.050	3.9E-10
Se-81	0.308h	F	0.800	8.6E-12	1.4E-11	0.800	2.7E-11
		M	0.800	1.5E-11	2.4E-11	0.050	2.7E-11
Se-81m	0.954h	F	0.800	1.7E-11	3.0E-11	0.800	5.3E-11
		M	0.800	4.7E-11	6.8E-11	0.050	5.9E-11
Se-83	0.375h	F	0.800	1.9E-11	3.4E-11	0.800	4.7E-11
		M	0.800	3.3E-11	5.3E-11	0.050	5.1E-11
Bromine							
Br-74	0.422h	F	1.000	2.8E-11	5.0E-11	1.000	8.4E-11
		M	1.000	4.1E-11	6.8E-11		
Br-74m	0.691h	F	1.000	4.2E-11	7.5E-11	1.000	1.4E-10
		M	1.000	6.5E-11	1.1E-10		
Br-75	1.63h	F	1.000	3.1E-11	5.6E-11	1.000	7.9E-11
		M	1.000	5.5E-11	8.5E-11		
Br-76	16.2h	F	1.000	2.6E-10	4.5E-10	1.000	4.6E-10
		M	1.000	4.2E-10	5.8E-10		
Br-77	2.33d	F	1.000	6.7E-11	1.2E-10	1.000	9.6E-11
		M	1.000	8.7E-11	1.3E-10		
Br-80	0.290h	F	1.000	6.3E-12	1.1E-11	1.000	3.1E-11
		M	1.000	1.0E-11	1.7E-11		
Br-80m	4.42h	F	1.000	3.5E-11	5.8E-11	1.000	1.1E-10
		M	1.000	7.6E-11	1.0E-10		

Table B.1.—(*continued*)

		Effective dose coefficients (Sv Bq^{-1})					
		Inhalation, $e_{inh}(50)$				Ingestion	
Nuclide	$t_{1/2}$	Type	f_1	1 μm AMAD	5 μm AMAD	f_1	$e_{ing}(50)$
Br-82	1.47d	F	1.000	3.7E-10	6.4E-10	1.000	5.4E-10
		M	1.000	6.4E-10	8.8E-10		
Br-83	2.39h	F	1.000	1.7E-11	2.9E-11	1.000	4.3E-11
		M	1.000	4.8E-11	6.7E-11		
Br-84	0.530h	F	1.000	2.3E-11	4.0E-11	1.000	8.8E-11
		M	1.000	3.9E-11	6.2E-11		
Rubidium							
Rb-79	0.382h	F	1.000	1.7E-11	3.0E-11	1.000	5.0E-11
Rb-81	4.58h	F	1.000	3.7E-11	6.8E-11	1.000	5.4E-11
Rb-81m	0.533h	F	1.000	7.3E-12	1.3E-11	1.000	9.7E-12
Rb-82m	6.20h	F	1.000	1.2E-10	2.2E-10	1.000	1.3E-10
Rb-83	86.2d	F	1.000	7.1E-10	1.0E-09	1.000	1.9E-09
Rb-84	32.8d	F	1.000	1.1E-09	1.5E-09	1.000	2.8E-09
Rb-86	18.6d	F	1.000	9.6E-10	1.3E-09	1.000	2.8E-09
Rb-87	4.70E+10y	F	1.000	5.1E-10	7.6E-10	1.000	1.5E-09
Rb-88	0.297h	F	1.000	1.7E-11	2.8E-11	1.000	9.0E-11
Rb-89	0.253h	F	1.000	1.4E-11	2.5E-11	1.000	4.7E-11
Strontium							
Sr-80	1.67h	F	0.300	7.6E-11	1.3E-10	0.300	3.4E-10
		S	0.010	1.4E-10	2.1E-10	0.010	3.5E-10
Sr-81	0.425h	F	0.300	2.2E-11	3.9E-11	0.300	7.7E-11
		S	0.010	3.8E-11	6.1E-11	0.010	7.8E-11
Sr-82	25.0d	F	0.300	2.2E-09	3.3E-09	0.300	6.1E-09
		S	0.010	1.0E-08	7.7E-09	0.010	6.0E-09
Sr-83	1.35d	F	0.300	1.7E-10	3.0E-10	0.300	4.9E-10
		S	0.010	3.4E-10	4.9E-10	0.010	5.8E-10
Sr-85	64.8d	F	0.300	3.9E-10	5.6E-10	0.300	5.6E-10
		S	0.010	7.7E-10	6.4E-10	0.010	3.3E-10

Table B.1.—(*continued*)

		Effective dose coefficients (Sv Bq^{-1})					
		Inhalation, $e_{inh}(50)$				Ingestion	
Nuclide	$t_{1/2}$	Type	f_1	1 μm AMAD	5 μm AMAD	f_1	$e_{ing}(50)$
Sr-85m	1.16h	F	0.300	3.1E-12	5.6E-12	0.300	6.1E-12
		S	0.010	4.5E-12	7.4E-12	0.010	6.1E-12
Sr-87m	2.80h	F	0.300	1.2E-11	2.2E-11	0.300	3.0E-11
		S	0.010	2.2E-11	3.5E-11	0.010	3.3E-11
Sr-89	50.5d	F	0.300	1.0E-09	1.4E-09	0.300	2.6E-09
		S	0.010	7.5E-09	5.6E-09	0.010	2.3E-09
Sr-90	29.1y	F	0.300	2.4E-08	3.0E-08	0.300	2.8E-08
		S	0.010	1.5E-07	7.7E-08	0.010	2.7E-09
Sr-91	9.50h	F	0.300	1.7E-10	2.9E-10	0.300	6.5E-10
		S	0.010	4.1E-10	5.7E-10	0.010	7.6E-10
Sr-92	2.71h	F	0.300	1.1E-10	1.8E-10	0.300	4.3E-10
		S	0.010	2.3E-10	3.4E-10	0.010	4.9E-10
Yttrium							
Y-86	14.7h	M	1.0E-04	4.8E-10	8.0E-10	1.0E-04	9.6E-10
		S	1.0E-04	4.9E-10	8.1E-10		
Y-86m	0.800h	M	1.0E-04	2.9E-11	4.8E-11	1.0E-04	5.6E-11
		S	1.0E-04	3.0E-11	4.9E-11		
Y-87	3.35d	M	1.0E-04	3.8E-10	5.2E-10	1.0E-04	5.5E-10
		S	1.0E-04	4.0E-10	5.3E-10		
Y-88	107d	M	1.0E-04	3.9E-09	3.3E-09	1.0E-04	1.3E-09
		S	1.0E-04	4.1E-09	3.0E-09		
Y-90	2.67d	M	1.0E-04	1.4E-09	1.6E-09	1.0E-04	2.7E-09
		S	1.0E-04	1.5E-09	1.7E-09		
Y-90m	3.19h	M	1.0E-04	9.6E-11	1.3E-10	1.0E-04	1.7E-10
		S	1.0E-04	1.0E-10	1.3E-10		
Y-91	58.5d	M	1.0E-04	6.7E-09	5.2E-09	1.0E-04	2.4E-09
		S	1.0E-04	8.4E-09	6.1E-09		
Y-91m	0.828h	M	1.0E-04	1.0E-11	1.4E-11	1.0E-04	1.1E-11
		S	1.0E-04	1.1E-11	1.5E-11		
Y-92	3.54h	M	1.0E-04	1.9E-10	2.7E-10	1.0E-04	4.9E-10
		S	1.0E-04	2.0E-10	2.8E-10		

Table B.1.—(*continued*)

		Effective dose coefficients (Sv Bq^{-1})					
		Inhalation, $e_{inh}(50)$				Ingestion	
Nuclide	$t_{1/2}$	Type	f_1	1 μm AMAD	5 μm AMAD	f_1	$e_{ing}(50)$
Y-93	10.1h	M	1.0E-04	4.1E-10	5.7E-10	1.0E-04	1.2E-09
		S	1.0E-04	4.3E-10	6.0E-10		
Y-94	0.318h	M	1.0E-04	2.8E-11	4.4E-11	1.0E-04	8.1E-11
		S	1.0E-04	2.9E-11	4.6E-11		
Y-95	0.178h	M	1.0E-04	1.6E-11	2.5E-11	1.0E-04	4.6E-11
		S	1.0E-04	1.7E-11	2.6E-11		
Zirconium							
Zr-86	16.5h	F	0.002	3.0E-10	5.2E-10	0.002	8.6E-10
		M	0.002	4.3E-10	6.8E-10		
		S	0.002	4.5E-10	7.0E-10		
Zr-88	83.4d	F	0.002	3.5E-09	4.1E-09	0.002	3.3E-10
		M	0.002	2.5E-09	1.7E-09		
		S	0.002	3.3E-09	1.8E-09		
Zr-89	3.27d	F	0.002	3.1E-10	5.2E-10	0.002	7.9E-10
		M	0.002	5.3E-10	7.2E-10		
		S	0.002	5.5E-10	7.5E-10		
Zr-93	1.53E+06y	F	0.002	2.5E-08	2.9E-08	0.002	2.8E-10
		M	0.002	9.6E-09	6.6E-09		
		S	0.002	3.1E-09	1.7E-09		
Zr-95	64.0d	F	0.002	2.5E-09	3.0E-09	0.002	8.8E-10
		M	0.002	4.5E-09	3.6E-09		
		S	0.002	5.5E-09	4.2E-09		
Zr-97	16.9h	F	0.002	4.2E-10	7.4E-10	0.002	2.1E-09
		M	0.002	9.4E-10	1.3E-09		
		S	0.002	1.0E-09	1.4E-09		
Niobium							
Nb-88	0.238h	M	0.010	2.9E-11	4.8E-11	0.010	6.3E-11
		S	0.010	3.0E-11	5.0E-11		
Nb-89	2.03h	M	0.010	1.2E-10	1.8E-10	0.010	3.0E-10
		S	0.010	1.3E-10	1.9E-10		
Nb-89	1.10h	M	0.010	7.1E-11	1.1E-10	0.010	1.4E-10
		S	0.010	7.4E-11	1.2E-10		
Nb-90	14.6h	M	0.010	6.6E-10	1.0E-09	0.010	1.2E-09
		S	0.010	6.9E-10	1.1E-09		

Table B.1.—(*continued*)

		Effective dose coefficients (Sv Bq^{-1})					
		Inhalation, $e_{inh}(50)$				Ingestion	
Nuclide	$t_{1/2}$	Type	f_1	1 μmAMAD	5 μmAMAD	f_1	$e_{ing}(50)$
Nb-93m	13.6y	M	0.010	4.6E-10	2.9E-10	0.010	1.2E-10
		S	0.010	1.6E-09	8.6E-10		
Nb-94	2.03E+04y	M	0.010	1.0E-08	7.2E-09	0.010	1.7E-09
		S	0.010	4.5E-08	2.5E-08		
Nb-95	35.1d	M	0.010	1.4E-09	1.3E-09	0.010	5.8E-10
		S	0.010	1.6E-09	1.3E-09		
Nb-95m	3.61d	M	0.010	7.6E-10	7.7E-10	0.010	5.6E-10
		S	0.010	8.5E-10	8.5E-10		
Nb-96	23.3h	M	0.010	6.5E-10	9.7E-10	0.010	1.1E-09
		S	0.010	6.8E-10	1.0E-09		
Nb-97	1.20h	M	0.010	4.4E-11	6.9E-11	0.010	6.8E-11
		S	0.010	4.7E-11	7.2E-11		
Nb-98	0.858h	M	0.010	5.9E-11	9.6E-11	0.010	1.1E-10
		S	0.010	6.1E-11	9.9E-11		
Molybdenum							
Mo-90	5.67h	F	0.800	1.7E-10	2.9E-10	0.800	3.1E-10
		S	0.050	3.7E-10	5.6E-10	0.050	6.2E-10
Mo-93	3.50E+03y	F	0.800	1.0E-09	1.4E-09	0.800	2.6E-09
		S	0.050	2.2E-09	1.2E-09	0.050	2.0E-10
Mo-93m	6.85h	F	0.800	1.0E-10	1.9E-10	0.800	1.6E-10
		S	0.050	1.8E-10	3.0E-10	0.050	2.8E-10
Mo-99	2.75d	F	0.800	2.3E-10	3.6E-10	0.800	7.4E-10
		S	0.050	9.7E-10	1.1E-09	0.050	1.2E-09
Mo-101	0.244h	F	0.800	1.5E-11	2.7E-11	0.800	4.2E-11
		S	0.050	2.7E-11	4.5E-11	0.050	4.2E-11
Technetium							
Tc-93	2.75h	F	0.800	3.4E-11	6.2E-11	0.800	4.9E-11
		M	0.800	3.6E-11	6.5E-11		
Tc-93m	0.725h	F	0.800	1.5E-11	2.6E-11	0.800	2.4E-11
		M	0.800	1.7E-11	3.1E-11		
Tc-94	4.88h	F	0.800	1.2E-10	2.1E-10	0.800	1.8E-10
		M	0.800	1.3E-10	2.2E-10		

Table B.1.—(*continued*)

		Effective dose coefficients (Sv Bq^{-1})					
		Inhalation, $e_{inh}(50)$				Ingestion	
Nuclide	$t_{1/2}$	Type	f_1	1 μm AMAD	5 μm AMAD	f_1	$e_{ing}(50)$
Tc-94m	0.867h	F	0.800	4.3E-11	6.9E-11	0.800	1.1E-10
		M	0.800	4.9E-11	8.0E-11		
Tc-95	20.0h	F	0.800	1.0E-10	1.8E-10	0.800	1.6E-10
		M	0.800	1.0E-10	1.8E-10		
Tc-95m	61.0d	F	0.800	3.1E-10	4.8E-10	0.800	6.2E-10
		M	0.800	8.7E-10	8.6E-10		
Tc-96	4.28d	F	0.800	6.0E-10	9.8E-10	0.800	1.1E-09
		M	0.800	7.1E-10	1.0E-09		
Tc-96m	0.858h	F	0.800	6.5E-12	1.1E-11	0.800	1.3E-11
		M	0.800	7.7E-12	1.1E-11		
Tc-97	2.60E+06y	F	0.800	4.5E-11	7.2E-11	0.800	8.3E-11
		M	0.800	2.1E-10	1.6E-10		
Tc-97m	87.0d	F	0.800	2.8E-10	4.0E-10	0.800	6.6E-10
		M	0.800	3.1E-09	2.7E-09		
Tc-98	4.20E+06y	F	0.800	1.0E-09	1.5E-09	0.800	2.3E-09
		M	0.800	8.1E-09	6.1E-09		
Tc-99	2.13E+05y	F	0.800	2.9E-10	4.0E-10	0.800	7.8E-10
		M	0.800	3.9E-09	3.2E-09		
Tc-99m	6.02h	F	0.800	1.2E-11	2.0E-11	0.800	2.2E-11
		M	0.800	1.9E-11	2.9E-11		
Tc-101	0.237h	F	0.800	8.7E-12	1.5E-11	0.800	1.9E-11
		M	0.800	1.3E-11	2.1E-11		
Tc-104	0.303h	F	0.800	2.4E-11	3.9E-11	0.800	8.1E-11
		M	0.800	3.0E-11	4.8E-11		
Ruthenium							
Ru-94	0.863h	F	0.050	2.7E-11	4.9E-11	0.050	9.4E-11
		M	0.050	4.4E-11	7.2E-11		
		S	0.050	4.6E-11	7.4E-11		
Ru-97	2.90d	F	0.050	6.7E-11	1.2E-10	0.050	1.5E-10
		M	0.050	1.1E-10	1.6E-10		
		S	0.050	1.1E-10	1.6E-10		

Table B.1.—(*continued*)

		Effective dose coefficients (Sv Bq^{-1})					
		Inhalation, $e_{inh}(50)$				Ingestion	
Nuclide	$t_{1/2}$	Type	f_1	1 μm AMAD	5 μm AMAD	f_1	$e_{ing}(50)$
Ru-103	39.3d	F	0.050	4.9E-10	6.8E-10	0.050	7.3E-10
		M	0.050	2.3E-09	1.9E-09		
		S	0.050	2.8E-09	2.2E-09		
Ru-105	4.44h	F	0.050	7.1E-11	1.3E-10	0.050	2.6E-10
		M	0.050	1.7E-10	2.4E-10		
		S	0.050	1.8E-10	2.5E-10		
Ru-106	1.01y	F	0.050	8.0E-09	9.8E-09	0.050	7.0E-09
		M	0.050	2.6E-08	1.7E-08		
		S	0.050	6.2E-08	3.5E-08		
Rhodium							
Rh-99	16.0d	F	0.050	3.3E-10	4.9E-10	0.050	5.1E-10
		M	0.050	7.3E-10	8.2E-10		
		S	0.050	8.3E-10	8.9E-10		
Rh-99m	4.70h	F	0.050	3.0E-11	5.7E-11	0.050	6.6E-11
		M	0.050	4.1E-11	7.2E-11		
		S	0.050	4.3E-11	7.3E-11		
Rh-100	20.8h	F	0.050	2.8E-10	5.1E-10	0.050	7.1E-10
		M	0.050	3.6E-10	6.2E-10		
		S	0.050	3.7E-10	6.3E-10		
Rh-101	3.20y	F	0.050	1.4E-09	1.7E-09	0.050	5.5E-10
		M	0.050	2.2E-09	1.7E-09		
		S	0.050	5.0E-09	3.1E-09		
Rh-101m	4.34d	F	0.050	1.0E-10	1.7E-10	0.050	2.2E-10
		M	0.050	2.0E-10	2.5E-10		
		S	0.050	2.1E-10	2.7E-10		
Rh-102	2.90y	F	0.050	7.3E-09	8.9E-09	0.050	2.6E-09
		M	0.050	6.5E-09	5.0E-09		
		S	0.050	1.6E-08	9.0E-09		
Rh-102m	207d	F	0.050	1.5E-09	1.9E-09	0.050	1.2E-09
		M	0.050	3.8E-09	2.7E-09		
		S	0.050	6.7E-09	4.2E-09		
Rh-103m	0.935h	F	0.050	8.6E-13	1.2E-12	0.050	3.8E-12
		M	0.050	2.3E-12	2.4E-12		
		S	0.050	2.5E-12	2.5E-12		

Table B.1.—(*continued*)

		Effective dose coefficients (Sv Bq^{-1})					
		Inhalation, $e_{inh}(50)$				Ingestion	
Nuclide	$t_{1/2}$	Type	f_1	1 μm AMAD	5 μm AMAD	f_1	$e_{ing}(50)$
Rh-105	1.47d	F	0.050	8.7E-11	1.5E-10	0.050	3.7E-10
		M	0.050	3.1E-10	4.1E-10		
		S	0.050	3.4E-10	4.4E-10		
Rh-106m	2.20h	F	0.050	7.0E-11	1.3E-10	0.050	1.6E-10
		M	0.050	1.1E-10	1.8E-10		
		S	0.050	1.2E-10	1.9E-10		
Rh-107	0.362h	F	0.050	9.6E-12	1.6E-11	0.050	2.4E-11
		M	0.050	1.7E-11	2.7E-11		
		S	0.050	1.7E-11	2.8E-11		
Palladium							
Pd-100	3.63d	F	0.005	4.9E-10	7.6E-10	0.005	9.4E-10
		M	0.005	7.9E-10	9.5E-10		
		S	0.005	8.3E-10	9.7E-10		
Pd-101	8.27h	F	0.005	4.2E-11	7.5E-11	0.005	9.4E-11
		M	0.005	6.2E-11	9.8E-11		
		S	0.005	6.4E-11	1.0E-10		
Pd-103	17.0d	F	0.005	9.0E-11	1.2E-10	0.005	1.9E-10
		M	0.005	3.5E-10	3.0E-10		
		S	0.005	4.0E-10	2.9E-10		
Pd-107	6.50E+06y	F	0.005	2.6E-11	3.3E-11	0.005	3.7E-11
		M	0.005	8.0E-11	5.2E-11		
		S	0.005	5.5E-10	2.9E-10		
Pd-109	13.4h	F	0.005	1.2E-10	2.1E-10	0.005	5.5E-10
		M	0.005	3.4E-10	4.7E-10		
		S	0.005	3.6E-10	5.0E-10		
Silver							
Ag-102	0.215h	F	0.050	1.4E-11	2.4E-11	0.050	4.0E-11
		M	0.050	1.8E-11	3.2E-11		
		S	0.050	1.9E-11	3.2E-11		
Ag-103	1.09h	F	0.050	1.6E-11	2.8E-11	0.050	4.3E-11
		M	0.050	2.7E-11	4.3E-11		
		S	0.050	2.8E-11	4.5E-11		
Ag-104	1.15h	F	0.050	3.0E-11	5.7E-11	0.050	6.0E-11
		M	0.050	3.9E-11	6.9E-11		
		S	0.050	4.0E-11	7.1E-11		

Table B.1.—(*continued*)

		Effective dose coefficients (Sv Bq^{-1})					
		Inhalation, $e_{inh}(50)$				Ingestion	
Nuclide	$t_{1/2}$	Type	f_1	1 μm AMAD	5 μm AMAD	f_1	$e_{ing}(50)$
Ag-104m	0.558h	F	0.050	1.7E-11	3.1E-11	0.050	5.4E-11
		M	0.050	2.6E-11	4.4E-11		
		S	0.050	2.7E-11	4.5E-11		
Ag-105	41.0d	F	0.050	5.4E-10	8.0E-10	0.050	4.7E-10
		M	0.050	6.9E-10	7.0E-10		
		S	0.050	7.8E-10	7.3E-10		
Ag-106	0.399h	F	0.050	9.8E-12	1.7E-11	0.050	3.2E-11
		M	0.050	1.6E-11	2.6E-11		
		S	0.050	1.6E-11	2.7E-11		
Ag-106m	8.41d	F	0.050	1.1E-09	1.6E-09	0.050	1.5E-09
		M	0.050	1.1E-09	1.5E-09		
		S	0.050	1.1E-09	1.4E-09		
Ag-108m	1.27E+02y	F	0.050	6.1E-09	7.3E-09	0.050	2.3E-09
		M	0.050	7.0E-09	5.2E-09		
		S	0.050	3.5E-08	1.9E-08		
Ag-110m	250d	F	0.050	5.5E-09	6.7E-09	0.050	2.8E-09
		M	0.050	7.2E-09	5.9E-09		
		S	0.050	1.2E-08	7.3E-09		
Ag-111	7.45d	F	0.050	4.1E-10	5.7E-10	0.050	1.3E-09
		M	0.050	1.5E-09	1.5E-09		
		S	0.050	1.7E-09	1.6E-09		
Ag-112	3.12h	F	0.050	8.2E-11	1.4E-10	0.050	4.3E-10
		M	0.050	1.7E-10	2.5E-10		
		S	0.050	1.8E-10	2.6E-10		
Ag-115	0.333h	F	0.050	1.6E-11	2.6E-11	0.050	6.0E-11
		M	0.050	2.8E-11	4.3E-11		
		S	0.050	3.0E-11	4.4E-11		
Cadmium							
Cd-104	0.961h	F	0.050	2.7E-11	5.0E-11	0.050	5.8E-11
		M	0.050	3.6E-11	6.2E-11		
		S	0.050	3.7E-11	6.3E-11		
Cd-107	6.49h	F	0.050	2.3E-11	4.2E-11	0.050	6.2E-11
		M	0.050	8.1E-11	1.0E-10		
		S	0.050	8.7E-11	1.1E-10		

Table B.1.—(*continued*)

		Effective dose coefficients (Sv Bq^{-1})					
		Inhalation, $e_{inh}(50)$				Ingestion	
Nuclide	$t_{1/2}$	Type	f_1	1 μm AMAD	5 μm AMAD	f_1	$e_{ing}(50)$
Cd-109	1.27y	F	0.050	8.1E-09	9.6E-09	0.050	2.0E-09
		M	0.050	6.2E-09	5.1E-09		
		S	0.050	5.8E-09	4.4E-09		
Cd-113	9.30E+15y	F	0.050	1.2E-07	1.4E-07	0.050	2.5E-08
		M	0.050	5.3E-08	4.3E-08		
		S	0.050	2.5E-08	2.1E-08		
Cd-113m	13.6y	F	0.050	1.1E-07	1.3E-07	0.050	2.3E-08
		M	0.050	5.0E-08	4.0E-08		
		S	0.050	3.0E-08	2.4E-08		
Cd-115	2.23d	F	0.050	3.7E-10	5.4E-10	0.050	1.4E-09
		M	0.050	9.7E-10	1.2E-09		
		S	0.050	1.1E-09	1.3E-09		
Cd-115m	44.6d	F	0.050	5.3E-09	6.4E-09	0.050	3.3E-09
		M	0.050	5.9E-09	5.5E-09		
		S	0.050	7.3E-09	5.5E-09		
Cd-117	2.49h	F	0.050	7.3E-11	1.3E-10	0.050	2.8E-10
		M	0.050	1.6E-10	2.4E-10		
		S	0.050	1.7E-10	2.5E-10		
Cd-117m	3.36h	F	0.050	1.0E-10	1.9E-10	0.050	2.8E-10
		M	0.050	2.0E-10	3.1E-10		
		S	0.050	2.1E-10	3.2E-10		
Indium							
In-109	4.20h	F	0.020	3.2E-11	5.7E-11	0.020	6.6E-11
		M	0.020	4.4E-11	7.3E-11		
In-110	4.90h	F	0.020	1.2E-10	2.2E-10	0.020	2.4E-10
		M	0.020	1.4E-10	2.5E-10		
In-110	1.15h	F	0.020	3.1E-11	5.5E-11	0.020	1.0E-10
		M	0.020	5.0E-11	8.1E-11		
In-111	2.83d	F	0.020	1.3E-10	2.2E-10	0.020	2.9E-10
		M	0.020	2.3E-10	3.1E-10		
In-112	0.240h	F	0.020	5.0E-12	8.6E-12	0.020	1.0E-11
		M	0.020	7.8E-12	1.3E-11		
In-113m	1.66h	F	0.020	1.0E-11	1.9E-11	0.020	2.8E-11
		M	0.020	2.0E-11	3.2E-11		

Table B.1.—(*continued*)

Nuclide	$t_{1/2}$	Type	f_1	Inhalation, $e_{inh}(50)$ 1 μm AMAD	5 μm AMAD	Ingestion f_1	$e_{ing}(50)$
In-114m	49.5d	F	0.020	9.3E-09	1.1E-08	0.020	4.1E-09
		M	0.020	5.9E-09	5.9E-09		
In-115	5.10E+15y	F	0.020	3.9E-07	4.5E-07	0.020	3.2E-08
		M	0.020	1.5E-07	1.1E-07		
In-115m	4.49h	F	0.020	2.5E-11	4.5E-11	0.020	8.6E-11
		M	0.020	6.0E-11	8.7E-11		
In-116m	0.902h	F	0.020	3.0E-11	5.5E-11	0.020	6.4E-11
		M	0.020	4.8E-11	8.0E-11		
In-117	0.730h	F	0.020	1.6E-11	2.8E-11	0.020	3.1E-11
		M	0.020	3.0E-11	4.8E-11		
In-117m	1.94h	F	0.020	3.1E-11	5.5E-11	0.020	1.2E-10
		M	0.020	7.3E-11	1.1E-10		
In-119m	0.300h	F	0.020	1.1E-11	1.8E-11	0.020	4.7E-11
		M	0.020	1.8E-11	2.9E-11		
Tin							
Sn-110	4.00h	F	0.020	1.1E-10	1.9E-10	0.020	3.5E-10
		M	0.020	1.6E-10	2.6E-10		
Sn-111	0.588h	F	0.020	8.3E-12	1.5E-11	0.020	2.3E-11
		M	0.020	1.4E-11	2.2E-11		
Sn-113	115d	F	0.020	5.4E-10	7.9E-10	0.020	7.3E-10
		M	0.020	2.5E-09	1.9E-09		
Sn-117m	13.6d	F	0.020	2.9E-10	3.9E-10	0.020	7.1E-10
		M	0.020	2.3E-09	2.2E-09		
Sn-119m	293d	F	0.020	2.9E-10	3.6E-10	0.020	3.4E-10
		M	0.020	2.0E-09	1.5E-09		
Sn-121	1.13d	F	0.020	6.4E-11	1.0E-10	0.020	2.3E-10
		M	0.020	2.2E-10	2.8E-10		
Sn-121m	55.0y	F	0.020	8.0E-10	9.7E-10	0.020	3.8E-10
		M	0.020	4.2E-09	3.3E-09		
Sn-123	129d	F	0.020	1.2E-09	1.6E-09	0.020	2.1E-09
		M	0.020	7.7E-09	5.6E-09		

Effective dose coefficients (Sv Bq^{-1})

Table B.1.—(*continued*)

				Effective dose coefficients (Sv Bq^{-1})			
		Inhalation, $e_{inh}(50)$				Ingestion	
Nuclide	$t_{1/2}$	Type	f_1	1 μm AMAD	5 μm AMAD	f_1	$e_{ing}(50)$
Sn-123m	0.668h	F	0.020	1.4E-11	2.4E-11	0.020	3.8E-11
		M	0.020	2.8E-11	4.4E-11		
Sn-125	9.64d	F	0.020	9.2E-10	1.3E-09	0.020	3.1E-09
		M	0.020	3.0E-09	2.8E-09		
Sn-126	1.00E+05y	F	0.020	1.1E-08	1.4E-08	0.020	4.7E-09
		M	0.020	2.7E-08	1.8E-08		
Sn-127	2.10h	F	0.020	6.9E-11	1.2E-10	0.020	2.0E-10
		M	0.020	1.3E-10	2.0E-10		
Sn-128	0.985h	F	0.020	5.4E-11	9.5E-11	0.020	1.5E-10
		M	0.020	9.6E-11	1.5E-10		
Antimony							
Sb-115	0.530h	F	0.100	9.2E-12	1.7E-11	0.100	2.4E-11
		M	0.010	1.4E-11	2.3E-11		
Sb-116	0.263h	F	0.100	9.9E-12	1.8E-11	0.100	2.6E-11
		M	0.010	1.4E-11	2.3E-11		
Sb-116m	1.00h	F	0.100	3.5E-11	6.4E-11	0.100	6.7E-11
		M	0.010	5.0E-11	8.5E-11		
Sb-117	2.80h	F	0.100	9.3E-12	1.7E-11	0.100	1.8E-11
		M	0.010	1.7E-11	2.7E-11		
Sb-118m	5.00h	F	0.100	1.0E-10	1.9E-10	0.100	2.1E-10
		M	0.010	1.3E-10	2.3E-10		
Sb-119	1.59d	F	0.100	2.5E-11	4.5E-11	0.100	8.1E-11
		M	0.010	3.7E-11	5.9E-11		
Sb-120	5.76d	F	0.100	5.9E-10	9.8E-10	0.100	1.2E-09
		M	0.010	1.0E-09	1.3E-09		
Sb-120	0.265h	F	0.100	4.9E-12	8.5E-12	0.100	1.4E-11
		M	0.010	7.4E-12	1.2E-11		
Sb-122	2.70d	F	0.100	3.9E-10	6.3E-10	0.100	1.7E-09
		M	0.010	1.0E-09	1.2E-09		
Sb-124	60.2d	F	0.100	1.3E-09	1.9E-09	0.100	2.5E-09
		M	0.010	6.1E-09	4.7E-09		

Table B.1.—(*continued*)

		Effective dose coefficients (Sv Bq^{-1})					
		Inhalation, $e_{inh}(50)$				Ingestion	
Nuclide	$t_{1/2}$	Type	f_1	1 μmAMAD	5 μmAMAD	f_1	$e_{ing}(50)$
Sb-124m	0.337h	F	0.100	3.0E-12	5.3E-12	0.100	8.0E-12
		M	0.010	5.5E-12	8.3E-12		
Sb-125	2.77y	F	0.100	1.4E-09	1.7E-09	0.100	1.1E-09
		M	0.010	4.5E-09	3.3E-09		
Sb-126	12.4d	F	0.100	1.1E-09	1.7E-09	0.100	2.4E-09
		M	0.010	2.7E-09	3.2E-09		
Sb-126m	0.317h	F	0.100	1.3E-11	2.3E-11	0.100	3.6E-11
		M	0.010	2.0E-11	3.3E-11		
Sb-127	3.85d	F	0.100	4.6E-10	7.4E-10	0.100	1.7E-09
		M	0.010	1.6E-09	1.7E-09		
Sb-128	9.01h	F	0.100	2.5E-10	4.6E-10	0.100	7.6E-10
		M	0.010	4.2E-10	6.7E-10		
Sb-128	0.173h	F	0.100	1.1E-11	1.9E-11	0.100	3.3E-11
		M	0.010	1.5E-11	2.6E-11		
Sb-129	4.32h	F	0.100	1.1E-10	2.0E-10	0.100	4.2E-10
		M	0.010	2.4E-10	3.5E-10		
Sb-130	0.667h	F	0.100	3.5E-11	6.3E-11	0.100	9.1E-11
		M	0.010	5.4E-11	9.1E-11		
Sb-131	0.383h	F	0.100	3.7E-11	5.9E-11	0.100	1.0E-10
		M	0.010	5.2E-11	8.3E-11		
Tellurium							
Te-116	2.49h	F	0.300	6.3E-11	1.2E-10	0.300	1.7E-10
		M	0.300	1.1E-10	1.7E-10		
Te-121	17.0d	F	0.300	2.5E-10	3.9E-10	0.300	4.3E-10
		M	0.300	3.9E-10	4.4E-10		
Te-121m	154d	F	0.300	1.8E-09	2.3E-09	0.300	2.3E-09
		M	0.300	4.2E-09	3.6E-09		
Te-123	1.00E+13y	F	0.300	4.0E-09	5.0E-09	0.300	4.4E-09
		M	0.300	2.6E-09	2.8E-09		
Te-123m	120d	F	0.300	9.7E-10	1.2E-09	0.300	1.4E-09
		M	0.300	3.9E-09	3.4E-09		

Table B.1.—(*continued*)

		Effective dose coefficients (Sv Bq^{-1})					
		Inhalation, $e_{inh}(50)$				Ingestion	
Nuclide	$t_{1/2}$	Type	f_1	1 μm AMAD	5 μm AMAD	f_1	$e_{ing}(50)$
Te-125m	58.0d	F	0.300	5.1E-10	6.7E-10	0.300	8.7E-10
		M	0.300	3.3E-09	2.9E-09		
Te-127	9.35h	F	0.300	4.2E-11	7.2E-11	0.300	1.7E-10
		M	0.300	1.2E-10	1.8E-10		
Te-127m	109d	F	0.300	1.6E-09	2.0E-09	0.300	2.3E-09
		M	0.300	7.2E-09	6.2E-09		
Te-129	1.16h	F	0.300	1.7E-11	2.9E-11	0.300	6.3E-11
		M	0.300	3.8E-11	5.7E-11		
Te-129m	33.6d	F	0.300	1.3E-09	1.8E-09	0.300	3.0E-09
		M	0.300	6.3E-09	5.4E-09		
Te-131	0.417h	F	0.300	2.3E-11	4.6E-11	0.300	8.7E-11
		M	0.300	3.8E-11	6.1E-11		
Te-131m	1.25d	F	0.300	8.7E-10	1.2E-09	0.300	1.9E-09
		M	0.300	1.1E-09	1.6E-09		
Te-132	3.26d	F	0.300	1.8E-09	2.4E-09	0.300	3.7E-09
		M	0.300	2.2E-09	3.0E-09		
Te-133	0.207h	F	0.300	2.0E-11	3.8E-11	0.300	7.2E-11
		M	0.300	2.7E-11	4.4E-11		
Te-133m	0.923h	F	0.300	8.4E-11	1.2E-10	0.300	2.8E-10
		M	0.300	1.2E-10	1.9E-10		
Te-134	0.696h	F	0.300	5.0E-11	8.3E-11	0.300	1.1E-10
		M	0.300	7.1E-11	1.1E-10		
Iodine							
I-120	1.35h	F	1.000	1.0E-10	1.9E-10	1.000	3.4E-10
I-120m	0.883h	F	1.000	8.7E-11	1.4E-10	1.000	2.1E-10
I-121	2.12h	F	1.000	2.8E-11	3.9E-11	1.000	8.2E-11
I-123	13.2h	F	1.000	7.6E-11	1.1E-10	1.000	2.1E-10
I-124	4.18d	F	1.000	4.5E-09	6.3E-09	1.000	1.3E-08
I-125	60.1d	F	1.000	5.3E-09	7.3E-09	1.000	1.5E-08
I-126	13.0d	F	1.000	1.0E-08	1.4E-08	1.000	2.9E-08

Table B.1.—(*continued*)

		Effective dose coefficients (Sv Bq^{-1})					
		Inhalation, $e_{inh}(50)$				Ingestion	
Nuclide	$t_{1/2}$	Type	f_1	1 μm AMAD	5 μm AMAD	f_1	$e_{ing}(50)$
I-128	0.416h	F	1.000	1.4E-11	2.2E-11	1.000	4.6E-11
I-129	1.57E+07y	F	1.000	3.7E-08	5.1E-08	1.000	1.1E-07
I-130	12.4h	F	1.000	6.9E-10	9.6E-10	1.000	2.0E-09
I-131	8.04d	F	1.000	7.6E-09	1.1E-08	1.000	2.2E-08
I-132	2.30h	F	1.000	9.6E-11	2.0E-10	1.000	2.9E-10
I-132m	1.39h	F	1.000	8.1E-11	1.1E-10	1.000	2.2E-10
I-133	20.8h	F	1.000	1.5E-09	2.1E-09	1.000	4.3E-09
I-134	0.876h	F	1.000	4.8E-11	7.9E-11	1.000	1.1E-10
I-135	6.61h	F	1.000	3.3E-10	4.6E-10	1.000	9.3E-10
Caesium							
Cs-125	0.750h	F	1.000	1.3E-11	2.3E-11	1.000	3.5E-11
Cs-127	6.25h	F	1.000	2.2E-11	4.0E-11	1.000	2.4E-11
Cs-129	1.34d	F	1.000	4.5E-11	8.1E-11	1.000	6.0E-11
Cs-130	0.498h	F	1.000	8.4E-12	1.5E-11	1.000	2.8E-11
Cs-131	9.69d	F	1.000	2.8E-11	4.5E-11	1.000	5.8E-11
Cs-132	6.48d	F	1.000	2.4E-10	3.8E-10	1.000	5.0E-10
Cs-134	2.06y	F	1.000	6.8E-09	9.6E-09	1.000	1.9E-08
Cs-134m	2.90h	F	1.000	1.5E-11	2.6E-11	1.000	2.0E-11
Cs-135	2.30E+06y	F	1.000	7.1E-10	9.9E-10	1.000	2.0E-09
Cs-135m	0.883h	F	1.000	1.3E-11	2.4E-11	1.000	1.9E-11
Cs-136	13.1d	F	1.000	1.3E-09	1.9E-09	1.000	3.0E-09
Cs-137	30.0y	F	1.000	4.8E-09	6.7E-09	1.000	1.3E-08
Cs-138	0.536h	F	1.000	2.6E-11	4.6E-11	1.000	9.2E-11

Table B.1.—(*continued*)

		Effective dose coefficients (Sv Bq^{-1})					
		Inhalation, $e_{inh}(50)$				Ingestion	
Nuclide	$t_{1/2}$	Type	f_1	1 μm AMAD	5 μm AMAD	f_1	$e_{ing}(50)$
Barium							
Ba-126	1.61h	F	0.100	7.8E-11	1.2E-10	0.100	2.6E-10
Ba-128	2.43d	F	0.100	8.0E-10	1.3E-09	0.100	2.7E-09
Ba-131	11.8d	F	0.100	2.3E-10	3.5E-10	0.100	4.5E-10
Ba-131m	0.243h	F	0.100	4.1E-12	6.4E-12	0.100	4.9E-12
Ba-133	10.7y	F	0.100	1.5E-09	1.8E-09	0.100	1.0E-09
Ba-133m	1.62d	F	0.100	1.9E-10	2.8E-10	0.100	5.5E-10
Ba-135m	1.20d	F	0.100	1.5E-10	2.3E-10	0.100	4.5E-10
Ba-139	1.38h	F	0.100	3.5E-11	5.5E-11	0.100	1.2E-10
Ba-140	12.7d	F	0.100	1.0E-09	1.6E-09	0.100	2.5E-09
Ba-141	0.305h	F	0.100	2.2E-11	3.5E-11	0.100	7.0E-11
Ba-142	0.177h	F	0.100	1.6E-11	2.7E-11	0.100	3.5E-11
Lanthanum							
La-131	0.983h	F	5.0E-04	1.4E-11	2.4E-11	5.0E-04	3.5E-11
		M	5.0E-04	2.3E-11	3.6E-11		
La-132	4.80h	F	5.0E-04	1.1E-10	2.0E-10	5.0E-04	3.9E-10
		M	5.0E-04	1.7E-10	2.8E-10		
La-135	19.5h	F	5.0E-04	1.1E-11	2.0E-11	5.0E-04	3.0E-11
		M	5.0E-04	1.5E-11	2.5E-11		
La-137	6.00E+04y	F	5.0E-04	8.6E-09	1.0E-08	5.0E-04	8.1E-11
		M	5.0E-04	3.4E-09	2.3E-09		
La-138	1.35E+11y	F	5.0E-04	1.5E-07	1.8E-07	5.0E-04	1.1E-09
		M	5.0E-04	6.1E-08	4.2E-08		
La-140	1.68d	F	5.0E-04	6.0E-10	1.0E-09	5.0E-04	2.0E-09
		M	5.0E-04	1.1E-09	1.5E-09		
La-141	3.93h	F	5.0E-04	6.7E-11	1.1E-10	5.0E-04	3.6E-10
		M	5.0E-04	1.5E-10	2.2E-10		

Table B.1.—(*continued*)

		Effective dose coefficients (Sv Bq^{-1})					
		Inhalation, $e_{inh}(50)$				Ingestion	
Nuclide	$t_{1/2}$	Type	f_1	1 μm AMAD	5 μm AMAD	f_1	$e_{ing}(50)$
La-142	1.54h	F	5.0E-04	5.6E-11	1.0E-10	5.0E-04	1.8E-10
		M	5.0E-04	9.3E-11	1.5E-10		
La-143	0.237h	F	5.0E-04	1.2E-11	2.0E-11	5.0E-04	5.6E-11
		M	5.0E-04	2.2E-11	3.3E-11		
Cerium							
Ce-134	3.00d	M	5.0E-04	1.3E-09	1.5E-09	5.0E-04	2.5E-09
		S	5.0E-04	1.3E-09	1.6E-09		
Ce-135	17.6h	M	5.0E-04	4.9E-10	7.3E-10	5.0E-04	7.9E-10
		S	5.0E-04	5.1E-10	7.6E-10		
Ce-137	9.00h	M	5.0E-04	1.0E-11	1.8E-11	5.0E-04	2.5E-11
		S	5.0E-04	1.1E-11	1.9E-11		
Ce-137m	1.43d	M	5.0E-04	4.0E-10	5.5E-10	5.0E-04	5.4E-10
		S	5.0E-04	4.3E-10	5.9E-10		
Ce-139	138d	M	5.0E-04	1.6E-09	1.3E-09	5.0E-04	2.6E-10
		S	5.0E-04	1.8E-09	1.4E-09		
Ce-141	32.5d	M	5.0E-04	3.1E-09	2.7E-09	5.0E-04	7.1E-10
		S	5.0E-04	3.6E-09	3.1E-09		
Ce-143	1.38d	M	5.0E-04	7.4E-10	9.5E-10	5.0E-04	1.1E-09
		S	5.0E-04	8.1E-10	1.0E-09		
Ce-144	284d	M	5.0E-04	3.4E-08	2.3E-08	5.0E-04	5.2E-09
		S	5.0E-04	4.9E-08	2.9E-08		
Praseodymium							
Pr-136	0.218h	M	5.0E-04	1.4E-11	2.4E-11	5.0E-04	3.3E-11
		S	5.0E-04	1.5E-11	2.5E-11		
Pr-137	1.28h	M	5.0E-04	2.1E-11	3.4E-11	5.0E-04	4.0E-11
		S	5.0E-04	2.2E-11	3.5E-11		
Pr-138m	2.10h	M	5.0E-04	7.6E-11	1.3E-10	5.0E-04	1.3E-10
		S	5.0E-04	7.9E-11	1.3E-10		
Pr-139	4.51h	M	5.0E-04	1.9E-11	2.9E-11	5.0E-04	3.1E-11
		S	5.0E-04	2.0E-11	3.0E-11		
Pr-142	19.1h	M	5.0E-04	5.3E-10	7.0E-10	5.0E-04	1.3E-09
		S	5.0E-04	5.6E-10	7.4E-10		

Table B.1.—(*continued*)

		Effective dose coefficients (Sv Bq^{-1})					
		Inhalation, $e_{inh}(50)$				Ingestion	
Nuclide	$t_{1/2}$	Type	f_1	1 μm AMAD	5 μm AMAD	f_1	$e_{ing}(50)$
Pr-142m	0.243h	M	5.0E-04	6.7E-12	8.9E-12	5.0E-04	1.7E-11
		S	5.0E-04	7.1E-12	9.4E-12		
Pr-143	13.6d	M	5.0E-04	2.1E-09	1.9E-09	5.0E-04	1.2E-09
		S	5.0E-04	2.3E-09	2.2E-09		
Pr-144	0.288h	M	5.0E-04	1.8E-11	2.9E-11	5.0E-04	5.0E-11
		S	5.0E-04	1.9E-11	3.0E-11		
Pr-145	5.98h	M	5.0E-04	1.6E-10	2.5E-10	5.0E-04	3.9E-10
		S	5.0E-04	1.7E-10	2.6E-10		
Pr-147	0.227h	M	5.0E-04	1.8E-11	2.9E-11	5.0E-04	3.3E-11
		S	5.0E-04	1.9E-11	3.0E-11		
Neodymium							
Nd-136	0.844h	M	5.0E-04	5.3E-11	8.5E-11	5.0E-04	9.9E-11
		S	5.0E-04	5.6E-11	8.9E-11		
Nd-138	5.04h	M	5.0E-04	2.4E-10	3.7E-10	5.0E-04	6.4E-10
		S	5.0E-04	2.6E-10	3.8E-10		
Nd-139	0.495h	M	5.0E-04	1.0E-11	1.7E-11	5.0E-04	2.0E-11
		S	5.0E-04	1.1E-11	1.7E-11		
Nd-139m	5.50h	M	5.0E-04	1.5E-10	2.5E-10	5.0E-04	2.5E-10
		S	5.0E-04	1.6E-10	2.5E-10		
Nd-141	2.49h	M	5.0E-04	5.1E-12	9.5E-12	5.0E-04	8.3E-12
		S	5.0E-04	5.3E-12	8.8E-12		
Nd-147	11.0d	M	5.0E-04	2.0E-09	1.9E-09	5.0E-04	1.1E-09
		S	5.0E-04	2.3E-09	2.1E-09		
Nd-149	1.73h	M	5.0E-04	8.5E-11	1.2E-10	5.0E-04	1.2E-10
		S	5.0E-04	9.0E-11	1.3E-10		
Nd-151	0.207h	M	5.0E-04	1.7E-11	2.8E-11	5.0E-04	3.0E-11
		S	5.0E-04	1.8E-11	2.9E-11		
Promethium							
Pm-141	0.348h	M	5.0E-04	1.5E-11	2.4E-11	5.0E-04	3.6E-11
		S	5.0E-04	1.6E-11	2.5E-11		
Pm-143	265d	M	5.0E-04	1.4E-09	9.6E-10	5.0E-04	2.3E-10
		S	5.0E-04	1.3E-09	8.3E-10		

Table B.1.—(*continued*)

		Effective dose coefficients (Sv Bq^{-1})					
		Inhalation, $e_{inh}(50)$				Ingestion	
Nuclide	$t_{1/2}$	Type	f_1	1 μm AMAD	5 μm AMAD	f_1	$e_{ing}(50)$
Pm-144	363d	M	5.0E-04	7.8E-09	5.4E-09	5.0E-04	9.7E-10
		S	5.0E-04	7.0E-09	3.9E-09		
Pm-145	17.7y	M	5.0E-04	3.4E-09	2.4E-09	5.0E-04	1.1E-10
		S	5.0E-04	2.1E-09	1.2E-09		
Pm-146	5.53y	M	5.0E-04	1.9E-08	1.3E-08	5.0E-04	9.0E-10
		S	5.0E-04	1.6E-08	9.0E-09		
Pm-147	2.62y	M	5.0E-04	4.7E-09	3.5E-09	5.0E-04	2.6E-10
		S	5.0E-04	4.6E-09	3.2E-09		
Pm-148	5.37d	M	5.0E-04	2.0E-09	2.1E-09	5.0E-04	2.7E-09
		S	5.0E-04	2.1E-09	2.2E-09		
Pm-148m	41.3d	M	5.0E-04	4.9E-09	4.1E-09	5.0E-04	1.8E-09
		S	5.0E-04	5.4E-09	4.3E-09		
Pm-149	2.21d	M	5.0E-04	6.6E-10	7.6E-10	5.0E-04	9.9E-10
		S	5.0E-04	7.2E-10	8.2E-10		
Pm-150	2.68h	M	5.0E-04	1.3E-10	2.0E-10	5.0E-04	2.6E-10
		S	5.0E-04	1.4E-10	2.1E-10		
Pm-151	1.18d	M	5.0E-04	4.2E-10	6.1E-10	5.0E-04	7.3E-10
		S	5.0E-04	4.5E-10	6.4E-10		
Samarium							
Sm-141	0.170h	M	5.0E-04	1.6E-11	2.7E-11	5.0E-04	3.9E-11
Sm-141m	0.377h	M	5.0E-04	3.4E-11	5.6E-11	5.0E-04	6.5E-11
Sm-142	1.21h	M	5.0E-04	7.4E-11	1.1E-10	5.0E-04	1.9E-10
Sm-145	340d	M	5.0E-04	1.5E-09	1.1E-09	5.0E-04	2.1E-10
Sm-146	1.03E+08y	M	5.0E-04	9.9E-06	6.7E-06	5.0E-04	5.4E-08
Sm-147	1.06E+11y	M	5.0E-04	8.9E-06	6.1E-06	5.0E-04	4.9E-08
Sm-151	90.0y	M	5.0E-04	3.7E-09	2.6E-09	5.0E-04	9.8E-11
Sm-153	1.95d	M	5.0E-04	6.1E-10	6.8E-10	5.0E-04	7.4E-10
Sm-155	0.368h	M	5.0E-04	1.7E-11	2.8E-11	5.0E-04	2.9E-11
Sm-156	9.40h	M	5.0E-04	2.1E-10	2.8E-10	5.0E-04	2.5E-10

Table B.1.—(*continued*)

		Effective dose coefficients (Sv Bq^{-1})					
		Inhalation, $e_{inh}(50)$				Ingestion	
Nuclide	$t_{1/2}$	Type	f_1	1 μm AMAD	5 μm AMAD	f_1	$e_{ing}(50)$
Europium							
Eu-145	5.94d	M	5.0E-04	5.6E-10	7.3E-10	5.0E-04	7.5E-10
Eu-146	4.61d	M	5.0E-04	8.2E-10	1.2E-09	5.0E-04	1.3E-09
Eu-147	24.0d	M	5.0E-04	1.0E-09	1.0E-09	5.0E-04	4.4E-10
Eu-148	54.5d	M	5.0E-04	2.7E-09	2.3E-09	5.0E-04	1.3E-09
Eu-149	93.1d	M	5.0E-04	2.7E-10	2.3E-10	5.0E-04	1.0E-10
Eu-150	34.2y	M	5.0E-04	5.0E-08	3.4E-08	5.0E-04	1.3E-09
Eu-150	12.6h	M	5.0E-04	1.9E-10	2.8E-10	5.0E-04	3.8E-10
Eu-152	13.3y	M	5.0E-04	3.9E-08	2.7E-08	5.0E-04	1.4E-09
Eu-152m	9.32h	M	5.0E-04	2.2E-10	3.2E-10	5.0E-04	5.0E-10
Eu-154	8.80y	M	5.0E-04	5.0E-08	3.5E-08	5.0E-04	2.0E-09
Eu-155	4.96y	M	5.0E-04	6.5E-09	4.7E-09	5.0E-04	3.2E-10
Eu-156	15.2d	M	5.0E-04	3.3E-09	3.0E-09	5.0E-04	2.2E-09
Eu-157	15.1h	M	5.0E-04	3.2E-10	4.4E-10	5.0E-04	6.0E-10
Eu-158	0.765h	M	5.0E-04	4.8E-11	7.5E-11	5.0E-04	9.4E-11
Gadolinium							
Gd-145	0.382h	F	5.0E-04	1.5E-11	2.6E-11	5.0E-04	4.4E-11
		M	5.0E-04	2.1E-11	3.5E-11		
Gd-146	48.3d	F	5.0E-04	4.4E-09	5.2E-09	5.0E-04	9.6E-10
		M	5.0E-04	6.0E-09	4.6E-09		
Gd-147	1.59d	F	5.0E-04	2.7E-10	4.5E-10	5.0E-04	6.1E-10
		M	5.0E-04	4.1E-10	5.9E-10		
Gd-148	93.0y	F	5.0E-04	2.5E-05	3.0E-05	5.0E-04	5.5E-08
		M	5.0E-04	1.1E-05	7.2E-06		
Gd-149	9.40d	F	5.0E-04	2.6E-10	4.5E-10	5.0E-04	4.5E-10
		M	5.0E-04	7.0E-10	7.9E-10		

Table B.1.—(*continued*)

		Effective dose coefficients (Sv Bq^{-1})					
		Inhalation, $e_{inh}(50)$				Ingestion	
Nuclide	$t_{1/2}$	Type	f_1	1 μm AMAD	5 μm AMAD	f_1	$e_{ing}(50)$
Gd-151	120d	F	5.0E-04	7.8E-10	9.3E-10	5.0E-04	2.0E-10
		M	5.0E-04	8.1E-10	6.5E-10		
Gd-152	1.08E+14y	F	5.0E-04	1.9E-05	2.2E-05	5.0E-04	4.1E-08
		M	5.0E-04	7.4E-06	5.0E-06		
Gd-153	242d	F	5.0E-04	2.1E-09	2.5E-09	5.0E-04	2.7E-10
		M	5.0E-04	1.9E-09	1.4E-09		
Gd-159	18.6h	F	5.0E-04	1.1E-10	1.8E-10	5.0E-04	4.9E-10
		M	5.0E-04	2.7E-10	3.9E-10		
Terbium							
Tb-147	1.65h	M	5.0E-04	7.9E-11	1.2E-10	5.0E-04	1.6E-10
Tb-149	4.15h	M	5.0E-04	4.3E-09	3.1E-09	5.0E-04	2.5E-10
Tb-150	3.27h	M	5.0E-04	1.1E-10	1.8E-10	5.0E-04	2.5E-10
Tb-151	17.6h	M	5.0E-04	2.3E-10	3.3E-10	5.0E-04	3.4E-10
Tb-153	2.34d	M	5.0E-04	2.0E-10	2.4E-10	5.0E-04	2.5E-10
Tb-154	21.4h	M	5.0E-04	3.8E-10	6.0E-10	5.0E-04	6.5E-10
Tb-155	5.32d	M	5.0E-04	2.1E-10	2.5E-10	5.0E-04	2.1E-10
Tb-156	5.34d	M	5.0E-04	1.2E-09	1.4E-09	5.0E-04	1.2E-09
Tb-156m	1.02d	M	5.0E-04	2.0E-10	2.3E-10	5.0E-04	1.7E-10
Tb-156m	5.00h	M	5.0E-04	9.2E-11	1.3E-10	5.0E-04	8.1E-11
Tb-157	1.50E+02y	M	5.0E-04	1.1E-09	7.9E-10	5.0E-04	3.4E-11
Tb-158	1.50E+02y	M	5.0E-04	4.3E-08	3.0E-08	5.0E-04	1.1E-09
Tb-160	72.3d	M	5.0E-04	6.6E-09	5.4E-09	5.0E-04	1.6E-09
Tb-161	6.91d	M	5.0E-04	1.2E-09	1.2E-09	5.0E-04	7.2E-10
Dysprosium							
Dy-155	10.0h	M	5.0E-04	8.0E-11	1.2E-10	5.0E-04	1.3E-10
Dy-157	8.10h	M	5.0E-04	3.2E-11	5.5E-11	5.0E-04	6.1E-11

Table B.1.—(*continued*)

			Effective dose coefficients (Sv Bq^{-1})				
			Inhalation, $e_{inh}(50)$			Ingestion	
Nuclide	$t_{1/2}$	Type	f_1	1 μm AMAD	5 μm AMAD	f_1	$e_{ing}(50)$
Dy-159	144d	M	5.0E-04	3.5E-10	2.5E-10	5.0E-04	1.0E-10
Dy-165	2.33h	M	5.0E-04	6.1E-11	8.7E-11	5.0E-04	1.1E-10
Dy-166	3.40d	M	5.0E-04	1.8E-09	1.8E-09	5.0E-04	1.6E-09
Holmium							
Ho-155	0.800h	M	5.0E-04	2.0E-11	3.2E-11	5.0E-04	3.7E-11
Ho-157	0.210h	M	5.0E-04	4.5E-12	7.6E-12	5.0E-04	6.5E-12
Ho-159	0.550h	M	5.0E-04	6.3E-12	1.0E-11	5.0E-04	7.9E-12
Ho-161	2.50h	M	5.0E-04	6.3E-12	1.0E-11	5.0E-04	1.3E-11
Ho-162	0.250h	M	5.0E-04	2.9E-12	4.5E-12	5.0E-04	3.3E-12
Ho-162m	1.13h	M	5.0E-04	2.2E-11	3.3E-11	5.0E-04	2.6E-11
Ho-164	0.483h	M	5.0E-04	8.6E-12	1.3E-11	5.0E-04	9.5E-12
Ho-164m	0.625h	M	5.0E-04	1.2E-11	1.6E-11	5.0E-04	1.6E-11
Ho-166	1.12d	M	5.0E-04	6.6E-10	8.3E-10	5.0E-04	1.4E-09
Ho-166m	1.20E+03y	M	5.0E-04	1.1E-07	7.8E-08	5.0E-04	2.0E-09
Ho-167	3.10h	M	5.0E-04	7.1E-11	1.0E-10	5.0E-04	8.3E-11
Erbium							
Er-161	3.24h	M	5.0E-04	5.1E-11	8.5E-11	5.0E-04	8.0E-11
Er-165	10.4h	M	5.0E-04	8.3E-12	1.4E-11	5.0E-04	1.9E-11
Er-169	9.30d	M	5.0E-04	9.8E-10	9.2E-10	5.0E-04	3.7E-10
Er-171	7.52h	M	5.0E-04	2.2E-10	3.0E-10	5.0E-04	3.6E-10
Er-172	2.05d	M	5.0E-04	1.1E-09	1.2E-09	5.0E-04	1.0E-09
Thulium							
Tm-162	0.362h	M	5.0E-04	1.6E-11	2.7E-11	5.0E-04	2.9E-11
Tm-166	7.70h	M	5.0E-04	1.8E-10	2.8E-10	5.0E-04	2.8E-10

Table B.1.—(*continued*)

		Effective dose coefficients (Sv Bq^{-1})					
		Inhalation, $e_{inh}(50)$				Ingestion	
Nuclide	$t_{1/2}$	Type	f_1	1 μm AMAD	5 μm AMAD	f_1	$e_{ing}(50)$
Tm-167	9.24d	M	5.0E-04	1.1E-09	1.0E-09	5.0E-04	5.6E-10
Tm-170	129d	M	5.0E-04	6.6E-09	5.2E-09	5.0E-04	1.3E-09
Tm-171	1.92y	M	5.0E-04	1.3E-09	9.1E-10	5.0E-04	1.1E-10
Tm-172	2.65d	M	5.0E-04	1.1E-09	1.4E-09	5.0E-04	1.7E-09
Tm-173	8.24h	M	5.0E-04	1.8E-10	2.6E-10	5.0E-04	3.1E-10
Tm-175	0.253h	M	5.0E-04	1.9E-11	3.1E-11	5.0E-04	2.7E-11
Ytterbium							
Yb-162	0.315h	M	5.0E-04	1.4E-11	2.2E-11	5.0E-04	2.3E-11
		S	5.0E-04	1.4E-11	2.3E-11		
Yb-166	2.36d	M	5.0E-04	7.2E-10	9.1E-10	5.0E-04	9.5E-10
		S	5.0E-04	7.6E-10	9.5E-10		
Yb-167	0.292h	M	5.0E-04	6.5E-12	9.0E-12	5.0E-04	6.7E-12
		S	5.0E-04	6.9E-12	9.5E-12		
Yb-169	32.0d	M	5.0E-04	2.4E-09	2.1E-09	5.0E-04	7.1E-10
		S	5.0E-04	2.8E-09	2.4E-09		
Yb-175	4.19d	M	5.0E-04	6.3E-10	6.4E-10	5.0E-04	4.4E-10
		S	5.0E-04	7.0E-10	7.0E-10		
Yb-177	1.90h	M	5.0E-04	6.4E-11	8.8E-11	5.0E-04	9.7E-11
		S	5.0E-04	6.9E-11	9.4E-11		
Yb-178	1.23h	M	5.0E-04	7.1E-11	1.0E-10	5.0E-04	1.2E-10
		S	5.0E-04	7.6E-11	1.1E-10		
Lutetium							
Lu-169	1.42d	M	5.0E-04	3.5E-10	4.7E-10	5.0E-04	4.6E-10
		S	5.0E-04	3.8E-10	4.9E-10		
Lu-170	2.00d	M	5.0E-04	6.4E-10	9.3E-10	5.0E-04	9.9E-10
		S	5.0E-04	6.7E-10	9.5E-10		
Lu-171	8.22d	M	5.0E-04	7.6E-10	8.8E-10	5.0E-04	6.7E-10
		S	5.0E-04	8.3E-10	9.3E-10		
Lu-172	6.70d	M	5.0E-04	1.4E-09	1.7E-09	5.0E-04	1.3E-09
		S	5.0E-04	1.5E-09	1.8E-09		

Table B.1.—(*continued*)

		Effective dose coefficients (Sv Bq^{-1})					
		Inhalation, $e_{inh}(50)$				Ingestion	
Nuclide	$t_{1/2}$	Type	f_1	1 μm AMAD	5 μm AMAD	f_1	$e_{ing}(50)$
Lu-173	1.37y	M	5.0E-04	2.0E-09	1.5E-09	5.0E-04	2.6E-10
		S	5.0E-04	2.3E-09	1.4E-09		
Lu-174	3.31y	M	5.0E-04	4.0E-09	2.9E-09	5.0E-04	2.7E-10
		S	5.0E-04	3.9E-09	2.5E-09		
Lu-174m	142d	M	5.0E-04	3.4E-09	2.4E-09	5.0E-04	5.3E-10
		S	5.0E-04	3.8E-09	2.6E-09		
Lu-176	3.60E+10y	M	5.0E-04	6.6E-08	4.6E-08	5.0E-04	1.8E-09
		S	5.0E-04	5.2E-08	3.0E-08		
Lu-176m	3.68h	M	5.0E-04	1.1E-10	1.5E-10	5.0E-04	1.7E-10
		S	5.0E-04	1.2E-10	1.6E-10		
Lu-177	6.71d	M	5.0E-04	1.0E-09	1.0E-09	5.0E-04	5.3E-10
		S	5.0E-04	1.1E-09	1.1E-09		
Lu-177m	161d	M	5.0E-04	1.2E-08	1.0E-08	5.0E-04	1.7E-09
		S	5.0E-04	1.5E-08	1.2E-08		
Lu-178	0.473h	M	5.0E-04	2.5E-11	3.9E-11	5.0E-04	4.7E-11
		S	5.0E-04	2.6E-11	4.1E-11		
Lu-178m	0.378h	M	5.0E-04	3.3E-11	5.4E-11	5.0E-04	3.8E-11
		S	5.0E-04	3.5E-11	5.6E-11		
Lu-179	4.59h	M	5.0E-04	1.1E-10	1.6E-10	5.0E-04	2.1E-10
		S	5.0E-04	1.2E-10	1.6E-10		
Hafnium							
Hf-170	16.0h	F	0.002	1.7E-10	2.9E-10	0.002	4.8E-10
		M	0.002	3.2E-10	4.3E-10		
Hf-172	1.87y	F	0.002	3.2E-08	3.7E-08	0.002	1.0E-09
		M	0.002	1.9E-08	1.3E-08		
Hf-173	24.0h	F	0.002	7.9E-11	1.3E-10	0.002	2.3E-10
		M	0.002	1.6E-10	2.2E-10		
Hf-175	70.0d	F	0.002	7.2E-10	8.7E-10	0.002	4.1E-10
		M	0.002	1.1E-09	8.8E-10		
Hf-177m	0.856h	F	0.002	4.7E-11	8.4E-11	0.002	8.1E-11
		M	0.002	9.2E-11	1.5E-10		

Table B.1.—(*continued*)

		Effective dose coefficients (Sv Bq^{-1})					
		Inhalation, $e_{inh}(50)$				Ingestion	
Nuclide	$t_{1/2}$	Type	f_1	1 μm AMAD	5 μm AMAD	f_1	$e_{ing}(50)$
Hf-178m	31.0y	F	0.002	2.6E-07	3.1E-07	0.002	4.7E-09
		M	0.002	1.1E-07	7.8E-08		
Hf-179m	25.1d	F	0.002	1.1E-09	1.4E-09	0.002	1.2E-09
		M	0.002	3.6E-09	3.2E-09		
Hf-180m	5.50h	F	0.002	6.4E-11	1.2E-10	0.002	1.7E-10
		M	0.002	1.4E-10	2.0E-10		
Hf-181	42.4d	F	0.002	1.4E-09	1.8E-09	0.002	1.1E-09
		M	0.002	4.7E-09	4.1E-09		
Hf-182	9.00E+06y	F	0.002	3.0E-07	3.6E-07	0.002	3.0E-09
		M	0.002	1.2E-07	8.3E-08		
Hf-182m	1.02h	F	0.002	2.3E-11	4.0E-11	0.002	4.2E-11
		M	0.002	4.7E-11	7.1E-11		
Hf-183	1.07h	F	0.002	2.6E-11	4.4E-11	0.002	7.3E-11
		M	0.002	5.8E-11	8.3E-11		
Hf-184	4.12h	F	0.002	1.3E-10	2.3E-10	0.002	5.2E-10
		M	0.002	3.3E-10	4.5E-10		
Tantalum							
Ta-172	0.613h	M	0.001	3.4E-11	5.5E-11	0.001	5.3E-11
		S	0.001	3.6E-11	5.7E-11		
Ta-173	3.65h	M	0.001	1.1E-10	1.6E-10	0.001	1.9E-10
		S	0.001	1.2E-10	1.6E-10		
Ta-174	1.20h	M	0.001	4.2E-11	6.3E-11	0.001	5.7E-11
		S	0.001	4.4E-11	6.6E-11		
Ta-175	10.5h	M	0.001	1.3E-10	2.0E-10	0.001	2.1E-10
		S	0.001	1.4E-10	2.0E-10		
Ta-176	8.08h	M	0.001	2.0E-10	3.2E-10	0.001	3.1E-10
		S	0.001	2.1E-10	3.3E-10		
Ta-177	2.36d	M	0.001	9.3E-11	1.2E-10	0.001	1.1E-10
		S	0.001	1.0E-10	1.3E-10		
Ta-178	2.20h	M	0.001	6.6E-11	1.0E-10	0.001	7.8E-11
		S	0.001	6.9E-11	1.1E-10		

Table B.1.—(*continued*)

		Effective dose coefficients (Sv Bq^{-1})					
		Inhalation, $e_{inh}(50)$				Ingestion	
Nuclide	$t_{1/2}$	Type	f_1	1 µm AMAD	5 µm AMAD	f_1	$e_{ing}(50)$
Ta-179	1.82y	M	0.001	2.0E-10	1.3E-10	0.001	6.5E-11
		S	0.001	5.2E-10	2.9E-10		
Ta-180	1.00E+13y	M	0.001	6.0E-09	4.6E-09	0.001	8.4E-10
		S	0.001	2.4E-08	1.4E-08		
Ta-180m	8.10h	M	0.001	4.4E-11	5.8E-11	0.001	5.4E-11
		S	0.001	4.7E-11	6.2E-11		
Ta-182	115d	M	0.001	7.2E-09	5.8E-09	0.001	1.5E-09
		S	0.001	9.7E-09	7.4E-09		
Ta-182m	0.264h	M	0.001	2.1E-11	3.4E-11	0.001	1.2E-11
		S	0.001	2.2E-11	3.6E-11		
Ta-183	5.10d	M	0.001	1.8E-09	1.8E-09	0.001	1.3E-09
		S	0.001	2.0E-09	2.0E-09		
Ta-184	8.70h	M	0.001	4.1E-10	6.0E-10	0.001	6.8E-10
		S	0.001	4.4E-10	6.3E-10		
Ta-185	0.816h	M	0.001	4.6E-11	6.8E-11	0.001	6.8E-11
		S	0.001	4.9E-11	7.2E-11		
Ta-186	0.175h	M	0.001	1.8E-11	3.0E-11	0.001	3.3E-11
		S	0.001	1.9E-11	3.1E-11		
Tungsten							
W-176	2.30h	F	0.300	4.4E-11	7.6E-11	0.300	1.0E-10
						0.010	1.1E-10
W-177	2.25h	F	0.300	2.6E-11	4.6E-11	0.300	5.8E-11
						0.010	6.1E-11
W-178	21.7d	F	0.300	7.6E-11	1.2E-10	0.300	2.2E-10
						0.010	2.5E-10
W-179	0.625h	F	0.300	9.9E-13	1.8E-12	0.300	3.3E-12
						0.010	3.3E-12
W-181	121d	F	0.300	2.8E-11	4.3E-11	0.300	7.6E-11
						0.010	8.2E-11
W-185	75.1d	F	0.300	1.4E-10	2.2E-10	0.300	4.4E-10
						0.010	5.0E-10

Table B.1.—(*continued*)

		Effective dose coefficients (Sv Bq^{-1})					
		Inhalation, $e_{inh}(50)$				Ingestion	
Nuclide	$t_{1/2}$	Type	f_1	1 μm AMAD	5 μm AMAD	f_1	$e_{ing}(50)$
W-187	23.9h	F	0.300	2.0E-10	3.3E-10	0.300	6.3E-10
						0.010	7.1E-10
W-188	69.4d	F	0.300	5.9E-10	8.4E-10	0.300	2.1E-09
						0.010	2.3E-09
Rhenium							
Re-177	0.233h	F	0.800	1.0E-11	1.7E-11	0.800	2.2E-11
		M	0.800	1.4E-11	2.2E-11		
Re-178	0.220h	F	0.800	1.1E-11	1.8E-11	0.800	2.5E-11
		M	0.800	1.5E-11	2.4E-11		
Re-181	20.0h	F	0.800	1.9E-10	3.0E-10	0.800	4.2E-10
		M	0.800	2.5E-10	3.7E-10		
Re-182	2.67d	F	0.800	6.8E-10	1.1E-09	0.800	1.4E-09
		M	0.800	1.3E-09	1.7E-09		
Re-182	12.7h	F	0.800	1.5E-10	2.4E-10	0.800	2.7E-10
		M	0.800	2.0E-10	3.0E-10		
Re-184	38.0d	F	0.800	4.6E-10	7.0E-10	0.800	1.0E-09
		M	0.800	1.8E-09	1.8E-09		
Re-184m	165d	F	0.800	6.1E-10	8.8E-10	0.800	1.5E-09
		M	0.800	6.1E-09	4.8E-09		
Re-186	3.78d	F	0.800	5.3E-10	7.3E-10	0.800	1.5E-09
		M	0.800	1.1E-09	1.2E-09		
Re-186m	2.00E+05y	F	0.800	8.5E-10	1.2E-09	0.800	2.2E-09
		M	0.800	1.1E-08	7.9E-09		
Re-187	5.00E+10y	F	0.800	1.9E-12	2.6E-12	0.800	5.1E-12
		M	0.800	6.0E-12	4.6E-12		
Re-188	17.0h	F	0.800	4.7E-10	6.6E-10	0.800	1.4E-09
		M	0.800	5.5E-10	7.4E-10		
Re-188m	0.310h	F	0.800	1.0E-11	1.6E-11	0.800	3.0E-11
		M	0.800	1.4E-11	2.0E-11		
Re-189	1.01d	F	0.800	2.7E-10	4.3E-10	0.800	7.8E-10
		M	0.800	4.3E-10	6.0E-10		

Table B.1.—(*continued*)

		Effective dose coefficients (Sv Bq^{-1})					
		Inhalation, $e_{inh}(50)$				Ingestion	
Nuclide	$t_{1/2}$	Type	f_1	1 μm AMAD	5 μm AMAD	f_1	$e_{ing}(50)$
Osmium							
Os-180	0.366h	F	0.010	8.8E-12	1.6E-11	0.010	1.7E-11
		M	0.010	1.4E-11	2.4E-11		
		S	0.010	1.5E-11	2.5E-11		
Os-181	1.75h	F	0.010	3.6E-11	6.4E-11	0.010	8.9E-11
		M	0.010	6.3E-11	9.6E-11		
		S	0.010	6.6E-11	1.0E-10		
Os-182	22.0h	F	0.010	1.9E-10	3.2E-10	0.010	5.6E-10
		M	0.010	3.7E-10	5.0E-10		
		S	0.010	3.9E-10	5.2E-10		
Os-185	94.0d	F	0.010	1.1E-09	1.4E-09	0.010	5.1E-10
		M	0.010	1.2E-09	1.0E-09		
		S	0.010	1.5E-09	1.1E-09		
Os-189m	6.00h	F	0.010	2.7E-12	5.2E-12	0.010	1.8E-11
		M	0.010	5.1E-12	7.6E-12		
		S	0.010	5.4E-12	7.9E-12		
Os-191	15.4d	F	0.010	2.5E-10	3.5E-10	0.010	5.7E-10
		M	0.010	1.5E-09	1.3E-09		
		S	0.010	1.8E-09	1.5E-09		
Os-191m	13.0h	F	0.010	2.6E-11	4.1E-11	0.010	9.6E-11
		M	0.010	1.3E-10	1.3E-10		
		S	0.010	1.5E-10	1.4E-10		
Os-193	1.25d	F	0.010	1.7E-10	2.8E-10	0.010	8.1E-10
		M	0.010	4.7E-10	6.4E-10		
		S	0.010	5.1E-10	6.8E-10		
Os-194	6.00y	F	0.010	1.1E-08	1.3E-08	0.010	2.4E-09
		M	0.010	2.0E-08	1.3E-08		
		S	0.010	7.9E-08	4.2E-08		
Iridium							
Ir-182	0.250h	F	0.010	1.5E-11	2.6E-11	0.010	4.8E-11
		M	0.010	2.4E-11	3.9E-11		
		S	0.010	2.5E-11	4.0E-11		
Ir-184	3.02h	F	0.010	6.7E-11	1.2E-10	0.010	1.7E-10
		M	0.010	1.1E-10	1.8E-10		
		S	0.010	1.2E-10	1.9E-10		

Table B.1.—(*continued*)

		Effective dose coefficients (Sv Bq^{-1})					
		Inhalation, $e_{inh}(50)$				Ingestion	
Nuclide	$t_{1/2}$	Type	f_1	1 μm AMAD	5 μm AMAD	f_1	$e_{ing}(50)$
Ir-185	14.0h	F	0.010	8.8E-11	1.5E-10	0.010	2.6E-10
		M	0.010	1.8E-10	2.5E-10		
		S	0.010	1.9E-10	2.6E-10		
Ir-186	15.8h	F	0.010	1.8E-10	3.3E-10	0.010	4.9E-10
		M	0.010	3.2E-10	4.8E-10		
		S	0.010	3.3E-10	5.0E-10		
Ir-186	1.75h	F	0.010	2.5E-11	4.5E-11	0.010	6.1E-11
		M	0.010	4.3E-11	6.9E-11		
		S	0.010	4.5E-11	7.1E-11		
Ir-187	10.5h	F	0.010	4.0E-11	7.2E-11	0.010	1.2E-10
		M	0.010	7.5E-11	1.1E-10		
		S	0.010	7.9E-11	1.2E-10		
Ir-188	1.73d	F	0.010	2.6E-10	4.4E-10	0.010	6.3E-10
		M	0.010	4.1E-10	6.0E-10		
		S	0.010	4.3E-10	6.2E-10		
Ir-189	13.3d	F	0.010	1.1E-10	1.7E-10	0.010	2.4E-10
		M	0.010	4.8E-10	4.1E-10		
		S	0.010	5.5E-10	4.6E-10		
Ir-190	12.1d	F	0.010	7.9E-10	1.2E-09	0.010	1.2E-09
		M	0.010	2.0E-09	2.3E-09		
		S	0.010	2.3E-09	2.5E-09		
Ir-190m	3.10h	F	0.010	5.3E-11	9.7E-11	0.010	1.2E-10
		M	0.010	8.3E-11	1.4E-10		
		S	0.010	8.6E-11	1.4E-10		
Ir-190m	1.20h	F	0.010	3.7E-12	5.6E-12	0.010	8.0E-12
		M	0.010	9.0E-12	1.0E-11		
		S	0.010	1.0E-11	1.1E-11		
Ir-192	74.0d	F	0.010	1.8E-09	2.2E-09	0.010	1.4E-09
		M	0.010	4.9E-09	4.1E-09		
		S	0.010	6.2E-09	4.9E-09		
Ir-192m	2.41E+02y	F	0.010	4.8E-09	5.6E-09	0.010	3.1E-10
		M	0.010	5.4E-09	3.4E-09		
		S	0.010	3.6E-08	1.9E-08		
Ir-193m	11.9d	F	0.010	1.0E-10	1.6E-10	0.010	2.7E-10
		M	0.010	1.0E-09	9.1E-10		
		S	0.010	1.2E-09	1.0E-09		

Table B.1.—(*continued*)

		Effective dose coefficients (Sv Bq^{-1})					
		Inhalation, $e_{inh}(50)$				Ingestion	
Nuclide	$t_{1/2}$	Type	f_1	1 μm AMAD	5 μm AMAD	f_1	$e_{ing}(50)$
Ir-194	19.1h	F	0.010	2.2E-10	3.6E-10	0.010	1.3E-09
		M	0.010	5.3E-10	7.1E-10		
		S	0.010	5.6E-10	7.5E-10		
Ir-194m	171d	F	0.010	5.4E-09	6.5E-09	0.010	2.1E-09
		M	0.010	8.5E-09	6.5E-09		
		S	0.010	1.2E-08	8.2E-09		
Ir-195	2.50h	F	0.010	2.6E-11	4.5E-11	0.010	1.0E-10
		M	0.010	6.7E-11	9.6E-11		
		S	0.010	7.2E-11	1.0E-10		
Ir-195m	3.80h	F	0.010	6.5E-11	1.1E-10	0.010	2.1E-10
		M	0.010	1.6E-10	2.3E-10		
		S	0.010	1.7E-10	2.4E-10		
Platinum							
Pt-186	2.00h	F	0.010	3.6E-11	6.6E-11	0.010	9.3E-11
Pt-188	10.2d	F	0.010	4.3E-10	6.3E-10	0.010	7.6E-10
Pt-189	10.9h	F	0.010	4.1E-11	7.3E-11	0.010	1.2E-10
Pt-191	2.80d	F	0.010	1.1E-10	1.9E-10	0.010	3.4E-10
Pt-193	50.0y	F	0.010	2.1E-11	2.7E-11	0.010	3.1E-11
Pt-193m	4.33d	F	0.010	1.3E-10	2.1E-10	0.010	4.5E-10
Pt-195m	4.02d	F	0.010	1.9E-10	3.1E-10	0.010	6.3E-10
Pt-197	18.3h	F	0.010	9.1E-11	1.6E-10	0.010	4.0E-10
Pt-197m	1.57h	F	0.010	2.5E-11	4.3E-11	0.010	8.4E-11
Pt-199	0.513h	F	0.010	1.3E-11	2.2E-11	0.010	3.9E-11
Pt-200	12.5h	F	0.010	2.4E-10	4.0E-10	0.010	1.2E-09
Gold							
Au-193	17.6h	F	0.100	3.9E-11	7.1E-11	0.100	1.3E-10
		M	0.100	1.1E-10	1.5E-10		
		S	0.100	1.2E-10	1.6E-10		

Table B.1.—(*continued*)

Nuclide	$t_{1/2}$	Effective dose coefficients (Sv Bq^{-1}) Inhalation, $e_{inh}(50)$				Ingestion	
		Type	f_1	1 µm AMAD	5 µm AMAD	f_1	$e_{ing}(50)$
Au-194	1.64d	F	0.100	1.5E-10	2.8E-10	0.100	4.2E-10
		M	0.100	2.4E-10	3.7E-10		
		S	0.100	2.5E-10	3.8E-10		
Au-195	183d	F	0.100	7.1E-11	1.2E-10	0.100	2.5E-10
		M	0.100	1.0E-09	8.0E-10		
		S	0.100	1.6E-09	1.2E-09		
Au-198	2.69d	F	0.100	2.3E-10	3.9E-10	0.100	1.0E-09
		M	0.100	7.6E-10	9.8E-10		
		S	0.100	8.4E-10	1.1E-09		
Au-198m	2.30d	F	0.100	3.4E-10	5.9E-10	0.100	1.3E-09
		M	0.100	1.7E-09	2.0E-09		
		S	0.100	1.9E-09	1.9E-09		
Au-199	3.14d	F	0.100	1.1E-10	1.9E-10	0.100	4.4E-10
		M	0.100	6.8E-10	6.8E-10		
		S	0.100	7.5E-10	7.6E-10		
Au-200	0.807h	F	0.100	1.7E-11	3.0E-11	0.100	6.8E-11
		M	0.100	3.5E-11	5.3E-11		
		S	0.100	3.6E-11	5.6E-11		
Au-200m	18.7h	F	0.100	3.2E-10	5.7E-10	0.100	1.1E-09
		M	0.100	6.9E-10	9.8E-10		
		S	0.100	7.3E-10	1.0E-09		
Au-201	0.440h	F	0.100	9.2E-12	1.6E-11	0.100	2.4E-11
		M	0.100	1.7E-11	2.8E-11		
		S	0.100	1.8E-11	2.9E-11		
Mercury							
Hg-193 (organic)	3.50h	F	0.400	2.6E-11	4.7E-11	1.000	3.1E-11
						0.400	6.6E-11
Hg-193 (inorganic)	3.50h	F	0.020	2.8E-11	5.0E-11	0.020	8.2E-11
		M	0.020	7.5E-11	1.0E-10		
Hg-193m (organic)	11.1h	F	0.400	1.1E-10	2.0E-10	1.000	1.3E-10
						0.400	3.0E-10
Hg-193m (inorganic)	11.1h	F	0.020	1.2E-10	2.3E-10	0.020	4.0E-10
		M	0.020	2.6E-10	3.8E-10		
Hg-194 (organic)	2.60E+02y	F	0.400	1.5E-08	1.9E-08	1.000	5.1E-08
						0.400	2.1E-08

Table B.1.—(*continued*)

		Effective dose coefficients (Sv Bq^{-1})					
		Inhalation, $e_{inh}(50)$				Ingestion	
Nuclide	$t_{1/2}$	Type	f_1	1 μm AMAD	5 μm AMAD	f_1	$e_{ing}(50)$
Hg-194 (inorganic)	2.60E+02y	F	0.020	1.3E-08	1.5E-08	0.020	1.4E-09
		M	0.020	7.8E-09	5.3E-09		
Hg-195 (organic)	9.90h	F	0.400	2.4E-11	4.4E-11	1.000	3.4E-11
						0.400	7.5E-11
Hg-195 (inorganic)	9.90h	F	0.020	2.7E-11	4.8E-11	0.020	9.7E-11
		M	0.020	7.2E-11	9.2E-11		
Hg-195m (organic)	1.73d	F	0.400	1.3E-10	2.2E-10	1.000	2.2E-10
						0.400	4.1E-10
Hg-195m (inorganic)	1.73d	F	0.020	1.5E-10	2.6E-10	0.020	5.6E-10
		M	0.020	5.1E-10	6.5E-10		
Hg-197 (organic)	2.67d	F	0.400	5.0E-11	8.5E-11	1.000	9.9E-11
						0.400	1.7E-10
Hg-197 (inorganic)	2.67d	F	0.020	6.0E-11	1.0E-10	0.020	2.3E-10
		M	0.020	2.9E-10	2.8E-10		
Hg-197m (organic)	23.8h	F	0.400	1.0E-10	1.8E-10	1.000	1.5E-10
						0.400	3.4E-10
Hg-197m (inorganic)	23.8h	F	0.020	1.2E-10	2.1E-10	0.020	4.7E-10
		M	0.020	5.1E-10	6.6E-10		
Hg-199m (organic)	0.710h	F	0.400	1.6E-11	2.7E-11	1.000	2.8E-11
						0.400	3.1E-11
Hg-199m (inorganic)	0.710h	F	0.020	1.6E-11	2.7E-11	0.020	3.1E-11
		M	0.020	3.3E-11	5.2E-11		
Hg-203 (organic)	46.6d	F	0.400	5.7E-10	7.5E-10	1.000	1.9E-09
						0.400	1.1E-09
Hg-203 (inorganic)	46.6d	F	0.020	4.7E-10	5.9E-10	0.020	5.4E-10
		M	0.020	2.3E-09	1.9E-09		
Thallium							
Tl-194	0.550h	F	1.000	4.8E-12	8.9E-12	1.000	8.1E-12
Tl-194m	0.546h	F	1.000	2.0E-11	3.6E-11	1.000	4.0E-11
Tl-195	1.16h	F	1.000	1.6E-11	3.0E-11	1.000	2.7E-11
Tl-197	2.84h	F	1.000	1.5E-11	2.7E-11	1.000	2.3E-11

Table B.1.—(*continued*)

		Effective dose coefficients (Sv Bq^{-1})					
		Inhalation, $e_{inh}(50)$				Ingestion	
Nuclide	$t_{1/2}$	Type	f_1	1 μm AMAD	5 μm AMAD	f_1	$e_{ing}(50)$
Tl-198	5.30h	F	1.000	6.6E-11	1.2E-10	1.000	7.3E-11
Tl-198m	1.87h	F	1.000	4.0E-11	7.3E-11	1.000	5.4E-11
Tl-199	7.42h	F	1.000	2.0E-11	3.7E-11	1.000	2.6E-11
Tl-200	1.09d	F	1.000	1.4E-10	2.5E-10	1.000	2.0E-10
Tl-201	3.04d	F	1.000	4.7E-11	7.6E-11	1.000	9.5E-11
Tl-202	12.2d	F	1.000	2.0E-10	3.1E-10	1.000	4.5E-10
Tl-204	3.78y	F	1.000	4.4E-10	6.2E-10	1.000	1.3E-09
Lead							
Pb-195m	0.263h	F	0.200	1.7E-11	3.0E-11	0.200	2.9E-11
Pb-198	2.40h	F	0.200	4.7E-11	8.7E-11	0.200	1.0E-10
Pb-199	1.50h	F	0.200	2.6E-11	4.8E-11	0.200	5.4E-11
Pb-200	21.5h	F	0.200	1.5E-10	2.6E-10	0.200	4.0E-10
Pb-201	9.40h	F	0.200	6.5E-11	1.2E-10	0.200	1.6E-10
Pb-202	3.00E+05y	F	0.200	1.1E-08	1.4E-08	0.200	8.7E-09
Pb-202m	3.62h	F	0.200	6.7E-11	1.2E-10	0.200	1.3E-10
Pb-203	2.17d	F	0.200	9.1E-11	1.6E-10	0.200	2.4E-10
Pb-205	1.43E+07y	F	0.200	3.4E-10	4.1E-10	0.200	2.8E-10
Pb-209	3.25h	F	0.200	1.8E-11	3.2E-11	0.200	5.7E-11
Pb-210	22.3y	F	0.200	8.9E-07	1.1E-06	0.200	6.8E-07
Pb-211	0.601h	F	0.200	3.9E-09	5.6E-09	0.200	1.8E-10
Pb-212	10.6h	F	0.200	1.9E-08	3.3E-08	0.200	5.9E-09
Pb-214	0.447h	F	0.200	2.9E-09	4.8E-09	0.200	1.4E-10
Bismuth							
Bi-200	0.606h	F	0.050	2.4E-11	4.2E-11	0.050	5.1E-11
		M	0.050	3.4E-11	5.6E-11		

Table B.1.—(*continued*)

		Effective dose coefficients (Sv Bq^{-1})					
		Inhalation, $e_{inh}(50)$				Ingestion	
Nuclide	$t_{1/2}$	Type	f_1	1 μm AMAD	5 μm AMAD	f_1	$e_{ing}(50)$
Bi-201	1.80h	F	0.050	4.7E-11	8.3E-11	0.050	1.2E-10
		M	0.050	7.0E-11	1.1E-10		
Bi-202	1.67h	F	0.050	4.6E-11	8.4E-11	0.050	8.9E-11
		M	0.050	5.8E-11	1.0E-10		
Bi-203	11.8h	F	0.050	2.0E-10	3.6E-10	0.050	4.8E-10
		M	0.050	2.8E-10	4.5E-10		
Bi-205	15.3d	F	0.050	4.0E-10	6.8E-10	0.050	9.0E-10
		M	0.050	9.2E-10	1.0E-09		
Bi-206	6.24d	F	0.050	7.9E-10	1.3E-09	0.050	1.9E-09
		M	0.050	1.7E-09	2.1E-09		
Bi-207	38.0y	F	0.050	5.2E-10	8.4E-10	0.050	1.3E-09
		M	0.050	5.2E-09	3.2E-09		
Bi-210	5.01d	F	0.050	1.1E-09	1.4E-09	0.050	1.3E-09
		M	0.050	8.4E-08	6.0E-08		
Bi-210m	3.00E+06y	F	0.050	4.5E-08	5.3E-08	0.050	1.5E-08
		M	0.050	3.1E-06	2.1E-06		
Bi-212	1.01h	F	0.050	9.3E-09	1.5E-08	0.050	2.6E-10
		M	0.050	3.0E-08	3.9E-08		
Bi-213	0.761h	F	0.050	1.1E-08	1.8E-08	0.050	2.0E-10
		M	0.050	2.9E-08	4.1E-08		
Bi-214	0.332h	F	0.050	7.2E-09	1.2E-08	0.050	1.1E-10
		M	0.050	1.4E-08	2.1E-08		
Polonium							
Po-203	0.612h	F	0.100	2.5E-11	4.5E-11	0.100	5.2E-11
		M	0.100	3.6E-11	6.1E-11		
Po-205	1.80h	F	0.100	3.5E-11	6.0E-11	0.100	5.9E-11
		M	0.100	6.4E-11	8.9E-11		
Po-207	5.83h	F	0.100	6.3E-11	1.2E-10	0.100	1.4E-10
		M	0.100	8.4E-11	1.5E-10		
Po-210	138d	F	0.100	6.0E-07	7.1E-07	0.100	2.4E-07
		M	0.100	3.0E-06	2.2E-06		

Table B.1.—(*continued*)

		Effective dose coefficients (Sv Bq^{-1})					
		Inhalation, $e_{inh}(50)$				Ingestion	
Nuclide	$t_{1/2}$	Type	f_1	1 μm AMAD	5 μm AMAD	f_1	$e_{ing}(50)$
Astatine							
At-207	1.80h	F	1.000	3.5E-10	4.4E-10	1.000	2.3E-10
		M	1.000	2.1E-09	1.9E-09		
At-211	7.21h	F	1.000	1.6E-08	2.7E-08	1.000	1.1E-08
		M	1.000	9.8E-08	1.1E-07		
Francium							
Fr-222	0.240h	F	1.000	1.4E-08	2.1E-08	1.000	7.1E-10
Fr-223	0.363h	F	1.000	9.1E-10	1.3E-09	1.000	2.3E-09
Radium							
Ra-223	11.4d	M	0.200	6.9E-06	5.7E-06	0.200	1.0E-07
Ra-224	3.66d	M	0.200	2.9E-06	2.4E-06	0.200	6.5E-08
Ra-225	14.8d	M	0.200	5.8E-06	4.8E-06	0.200	9.5E-08
Ra-226	1.60E+03y	M	0.200	1.6E-05	1.2E-05	0.200	2.8E-07
Ra-227	0.703h	M	0.200	2.8E-10	2.1E-10	0.200	8.4E-11
Ra-228	5.75y	M	0.200	2.6E-06	1.7E-06	0.200	6.7E-07
Actinium							
Ac-224	2.90h	F	5.0E-04	1.1E-08	1.3E-08	5.0E-04	7.0E-10
		M	5.0E-04	1.0E-07	8.9E-08		
		S	5.0E-04	1.2E-07	9.9E-08		
Ac-225	10.0d	F	5.0E-04	8.7E-07	1.0E-06	5.0E-04	2.4E-08
		M	5.0E-04	6.9E-06	5.7E-06		
		S	5.0E-04	7.9E-06	6.5E-06		
Ac-226	1.21d	F	5.0E-04	9.5E-08	2.2E-07	5.0E-04	1.0E-08
		M	5.0E-04	1.1E-06	9.2E-07		
		S	5.0E-04	1.2E-06	1.0E-06		
Ac-227	21.8y	F	5.0E-04	5.4E-04	6.3E-04	5.0E-04	1.1E-06
		M	5.0E-04	2.1E-04	1.5E-04		
		S	5.0E-04	6.6E-05	4.7E-05		

Table B.1.—(*continued*)

		Effective dose coefficients (Sv Bq^{-1})					
		Inhalation, $e_{inh}(50)$				Ingestion	
Nuclide	$t_{1/2}$	Type	f_1	1 μm AMAD	5 μm AMAD	f_1	$e_{ing}(50)$
Ac-228	6.13h	F	5.0E-04	2.5E-08	2.9E-08	5.0E-04	4.3E-10
		M	5.0E-04	1.6E-08	1.2E-08		
		S	5.0E-04	1.4E-08	1.2E-08		
Thorium							
Th-226	0.515h	M	5.0E-04	5.5E-08	7.4E-08	5.0E-04	3.5E-10
		S	2.0E-04	5.9E-08	7.8E-08	2.0E-04	3.6E-10
Th-227	18.7d	M	5.0E-04	7.8E-06	6.2E-06	5.0E-04	8.9E-09
		S	2.0E-04	9.6E-06	7.6E-06	2.0E-04	8.4E-09
Th-228	1.91y	M	5.0E-04	3.1E-05	2.3E-05	5.0E-04	7.0E-08
		S	2.0E-04	3.9E-05	3.2E-05	2.0E-04	3.5E-08
Th-229	7.34E+03y	M	5.0E-04	9.9E-05	6.9E-05	5.0E-04	4.8E-07
		S	2.0E-04	6.5E-05	4.8E-05	2.0E-04	2.0E-07
Th-230	7.70E+04y	M	5.0E-04	4.0E-05	2.8E-05	5.0E-04	2.1E-07
		S	2.0E-04	1.3E-05	7.2E-06	2.0E-04	8.7E-08
Th-231	1.06d	M	5.0E-04	2.9E-10	3.7E-10	5.0E-04	3.4E-10
		S	2.0E-04	3.2E-10	4.0E-10	2.0E-04	3.4E-10
Th-232	1.40E+10y	M	5.0E-04	4.2E-05	2.9E-05	5.0E-04	2.2E-07
		S	2.0E-04	2.3E-05	1.2E-05	2.0E-04	9.2E-08
Th-234	24.1d	M	5.0E-04	6.3E-09	5.3E-09	5.0E-04	3.4E-09
		S	2.0E-04	7.3E-09	5.8E-09	2.0E-04	3.4E-09
Protactinium							
Pa-227	0.638h	M	5.0E-04	7.0E-08	9.0E-08	5.0E-04	4.5E-10
		S	5.0E-04	7.6E-08	9.7E-08		
Pa-228	22.0h	M	5.0E-04	5.9E-08	4.6E-08	5.0E-04	7.8E-10
		S	5.0E-04	6.9E-08	5.1E-08		
Pa-230	17.4d	M	5.0E-04	5.6E-07	4.6E-07	5.0E-04	9.2E-10
		S	5.0E-04	7.1E-07	5.7E-07		
Pa-231	3.27E+04y	M	5.0E-04	1.3E-04	8.9E-05	5.0E-04	7.1E-07
		S	5.0E-04	3.2E-05	1.7E-05		
Pa-232	1.31d	M	5.0E-04	9.5E-09	6.8E-09	5.0E-04	7.2E-10
		S	5.0E-04	3.2E-09	2.0E-09		

Table B.1.—(*continued*)

		Effective dose coefficients (Sv Bq^{-1})					
		Inhalation, $e_{inh}(50)$				Ingestion	
Nuclide	$t_{1/2}$	Type	f_1	1 μm AMAD	5 μm AMAD	f_1	$e_{ing}(50)$
Pa-233	27.0d	M	5.0E-04	3.1E-09	2.8E-09	5.0E-04	8.7E-10
		S	5.0E-04	3.7E-09	3.2E-09		
Pa-234	6.70h	M	5.0E-04	3.8E-10	5.5E-10	5.0E-04	5.1E-10
		S	5.0E-04	4.0E-10	5.8E-10		
Uranium							
U-230	20.8d	F	0.020	3.6E-07	4.2E-07	0.020	5.5E-08
		M	0.020	1.2E-05	1.0E-05	0.002	2.8E-08
		S	0.002	1.5E-05	1.2E-05		
U-231	4.20d	F	0.020	8.3E-11	1.4E-10	0.020	2.8E-10
		M	0.020	3.4E-10	3.7E-10	0.002	2.8E-10
		S	0.002	3.7E-10	4.0E-10		
U-232	72.0y	F	0.020	4.0E-06	4.7E-06	0.020	3.3E-07
		M	0.020	7.2E-06	4.8E-06	0.002	3.7E-08
		S	0.002	3.5E-05	2.6E-05		
U-233	1.58E+05y	F	0.020	5.7E-07	6.6E-07	0.020	5.0E-08
		M	0.020	3.2E-06	2.2E-06	0.002	8.5E-09
		S	0.002	8.7E-06	6.9E-06		
U-234	2.44E+05y	F	0.020	5.5E-07	6.4E-07	0.020	4.9E-08
		M	0.020	3.1E-06	2.1E-06	0.002	8.3E-09
		S	0.002	8.5E-06	6.8E-06		
U-235	7.04E+08y	F	0.020	5.1E-07	6.0E-07	0.020	4.6E-08
		M	0.020	2.8E-06	1.8E-06	0.002	8.3E-09
		S	0.002	7.7E-06	6.1E-06		
U-236	2.34E+07y	F	0.020	5.2E-07	6.1E-07	0.020	4.6E-08
		M	0.020	2.9E-06	1.9E-06	0.002	7.9E-09
		S	0.002	7.9E-06	6.3E-06		
U-237	6.75d	F	0.020	1.9E-10	3.3E-10	0.020	7.6E-10
		M	0.020	1.6E-09	1.5E-09	0.002	7.7E-10
		S	0.002	1.8E-09	1.7E-09		
U-238	4.47E+09y	F	0.020	4.9E-07	5.8E-07	0.020	4.4E-08
		M	0.020	2.6E-06	1.6E-06	0.002	7.6E-09
		S	0.002	7.3E-06	5.7E-06		
U-239	0.392h	F	0.020	1.1E-11	1.8E-11	0.020	2.7E-11
		M	0.020	2.3E-11	3.3E-11	0.002	2.8E-11
		S	0.002	2.4E-11	3.5E-11		

Table B.1.—(*continued*)

		Effective dose coefficients (Sv Bq^{-1})					
		Inhalation, $e_{inh}(50)$				Ingestion	
Nuclide	$t_{1/2}$	Type	f_1	1 μm AMAD	5 μm AMAD	f_1	$e_{ing}(50)$
U-240	14.1h	F	0.020	2.1E-10	3.7E-10	0.020	1.1E-09
		M	0.020	5.3E-10	7.9E-10	0.002	1.1E-09
		S	0.002	5.7E-10	8.4E-10		
Neptunium							
Np-232	0.245h	M	5.0E-04	4.7E-11	3.5E-11	5.0E-04	9.7E-12
Np-233	0.603h	M	5.0E-04	1.7E-12	3.0E-12	5.0E-04	2.2E-12
Np-234	4.40d	M	5.0E-04	5.4E-10	7.3E-10	5.0E-04	8.1E-10
Np-235	1.08y	M	5.0E-04	4.0E-10	2.7E-10	5.0E-04	5.3E-11
Np-236	1.15E+05y	M	5.0E-04	3.0E-06	2.0E-06	5.0E-04	1.7E-08
Np-236	22.5h	M	5.0E-04	5.0E-09	3.6E-09	5.0E-04	1.9E-10
Np-237	2.14E+06y	M	5.0E-04	2.1E-05	1.5E-05	5.0E-04	1.1E-07
Np-238	2.12d	M	5.0E-04	2.0E-09	1.7E-09	5.0E-04	9.1E-10
Np-239	2.36d	M	5.0E-04	9.0E-10	1.1E-09	5.0E-04	8.0E-10
Np-240	1.08h	M	5.0E-04	8.7E-11	1.3E-10	5.0E-04	8.2E-11
Plutonium							
Pu-234	8.80h	M	5.0E-04	1.9E-08	1.6E-08	5.0E-04	1.6E-10
		S	1.0E-05	2.2E-08	1.8E-08	1.0E-05	1.5E-10
						1.0E-04	1.6E-10
Pu-235	0.422h	M	5.0E-04	1.5E-12	2.5E-12	5.0E-04	2.1E-12
		S	1.0E-05	1.6E-12	2.6E-12	1.0E-05	2.1E-12
						1.0E-04	2.1E-12
Pu-236	2.85y	M	5.0E-04	1.8E-05	1.3E-05	5.0E-04	8.6E-08
		S	1.0E-05	9.6E-06	7.4E-06	1.0E-05	6.3E-09
						1.0E-04	2.1E-08
Pu-237	45.3d	M	5.0E-04	3.3E-10	2.9E-10	5.0E-04	1.0E-10
		S	1.0E-05	3.6E-10	3.0E-10	1.0E-05	1.0E-10
						1.0E-04	1.0E-10
Pu-238	87.7y	M	5.0E-04	4.3E-05	3.0E-05	5.0E-04	2.3E-07
		S	1.0E-05	1.5E-05	1.1E-05	1.0E-05	8.8E-09
						1.0E-04	4.9E-08

Table B.1.—(*continued*)

		Effective dose coefficients (Sv Bq^{-1})					
		Inhalation, $e_{inh}(50)$				Ingestion	
Nuclide	$t_{1/2}$	Type	f_1	1 μm AMAD	5 μm AMAD	f_1	$e_{ing}(50)$
Pu-239	2.41E+04y	M	5.0E-04	4.7E-05	3.2E-05	5.0E-04	2.5E-07
		S	1.0E-05	1.5E-05	8.3E-06	1.0E-05	9.0E-09
						1.0E-04	5.3E-08
Pu-240	6.54E+03y	M	5.0E-04	4.7E-05	3.2E-05	5.0E-04	2.5E-07
		S	1.0E-05	1.5E-05	8.3E-06	1.0E-05	9.0E-09
						1.0E-04	5.3E-08
Pu-241	14.4y	M	5.0E-04	8.5E-07	5.8E-07	5.0E-04	4.7E-09
		S	1.0E-05	1.6E-07	8.4E-08	1.0E-05	1.1E-10
						1.0E-04	9.6E-10
Pu-242	3.76E+05y	M	5.0E-04	4.4E-05	3.1E-05	5.0E-04	2.4E-07
		S	1.0E-05	1.4E-05	7.7E-06	1.0E-05	8.6E-09
						1.0E-04	5.0E-08
Pu-243	4.95h	M	5.0E-04	8.2E-11	1.1E-10	5.0E-04	8.5E-11
		S	1.0E-05	8.5E-11	1.1E-10	1.0E-05	8.5E-11
						1.0E-04	8.5E-11
Pu-244	8.26E+07y	M	5.0E-04	4.4E-05	3.0E-05	5.0E-04	2.4E-07
		S	1.0E-05	1.3E-05	7.4E-06	1.0E-05	1.1E-08
						1.0E-04	5.2E-08
Pu-245	10.5h	M	5.0E-04	4.5E-10	6.1E-10	5.0E-04	7.2E-10
		S	1.0E-05	4.8E-10	6.5E-10	1.0E-05	7.2E-10
						1.0E-04	7.2E-10
Pu-246	10.9d	M	5.0E-04	7.0E-09	6.5E-09	5.0E-04	3.3E-09
		S	1.0E-05	7.6E-09	7.0E-09	1.0E-05	3.3E-09
						1.0E-04	3.3E-09
Americium							
Am-237	1.22h	M	5.0E-04	2.5E-11	3.6E-11	5.0E-04	1.8E-11
Am-238	1.63h	M	5.0E-04	8.5E-11	6.6E-11	5.0E-04	3.2E-11
Am-239	11.9h	M	5.0E-04	2.2E-10	2.9E-10	5.0E-04	2.4E-10
Am-240	2.12d	M	5.0E-04	4.4E-10	5.9E-10	5.0E-04	5.8E-10
Am-241	4.32E+02y	M	5.0E-04	3.9E-05	2.7E-05	5.0E-04	2.0E-07
Am-242	16.0h	M	5.0E-04	1.6E-08	1.2E-08	5.0E-04	3.0E-10
Am-242m	1.52E+02y	M	5.0E-04	3.5E-05	2.4E-05	5.0E-04	1.9E-07

Table B.1.—(*continued*)

		Effective dose coefficients (Sv Bq^{-1})						
		Inhalation, $e_{inh}(50)$					Ingestion	
Nuclide	$t_{1/2}$	Type	f_1	1 μm AMAD	5 μm AMAD		f_1	$e_{ing}(50)$
Am-243	7.38E+03y	M	5.0E-04	3.9E-05	2.7E-05		5.0E-04	2.0E-07
Am-244	10.1h	M	5.0E-04	1.9E-09	1.5E-09		5.0E-04	4.6E-10
Am-244m	0.433h	M	5.0E-04	7.9E-11	6.2E-11		5.0E-04	2.9E-11
Am-245	2.05h	M	5.0E-04	5.3E-11	7.6E-11		5.0E-04	6.2E-11
Am-246	0.650h	M	5.0E-04	6.8E-11	1.1E-10		5.0E-04	5.8E-11
Am-246m	0.417h	M	5.0E-04	2.3E-11	3.8E-11		5.0E-04	3.4E-11
Curium								
Cm-238	2.40h	M	5.0E-04	4.1E-09	4.8E-09		5.0E-04	8.0E-11
Cm-240	27.0d	M	5.0E-04	2.9E-06	2.3E-06		5.0E-04	7.6E-09
Cm-241	32.8d	M	5.0E-04	3.4E-08	2.6E-08		5.0E-04	9.1E-10
Cm-242	163d	M	5.0E-04	4.8E-06	3.7E-06		5.0E-04	1.2E-08
Cm-243	28.5y	M	5.0E-04	2.9E-05	2.0E-05		5.0E-04	1.5E-07
Cm-244	18.1y	M	5.0E-04	2.5E-05	1.7E-05		5.0E-04	1.2E-07
Cm-245	8.50E+03y	M	5.0E-04	4.0E-05	2.7E-05		5.0E-04	2.1E-07
Cm-246	4.73E+03y	M	5.0E-04	4.0E-05	2.7E-05		5.0E-04	2.1E-07
Cm-247	1.56E+07y	M	5.0E-04	3.6E-05	2.5E-05		5.0E-04	1.9E-07
Cm-248	3.39E+05y	M	5.0E-04	1.4E-04	9.5E-05		5.0E-04	7.7E-07
Cm-249	1.07h	M	5.0E-04	3.2E-11	5.1E-11		5.0E-04	3.1E-11
Cm-250	6.90E+03y	M	5.0E-04	7.9E-04	5.4E-04		5.0E-04	4.4E-06
Berkelium								
Bk-245	4.94d	M	5.0E-04	2.0E-09	1.8E-09		5.0E-04	5.7E-10
Bk-246	1.83d	M	5.0E-04	3.4E-10	4.6E-10		5.0E-04	4.8E-10
Bk-247	1.38E+03y	M	5.0E-04	6.5E-05	4.5E-05		5.0E-04	3.5E-07
Bk-249	320d	M	5.0E-04	1.5E-07	1.0E-07		5.0E-04	9.7E-10

Table B.1.—(*continued*)

		Effective dose coefficients (Sv Bq^{-1})					
		Inhalation, $e_{inh}(50)$				Ingestion	
Nuclide	$t_{1/2}$	Type	f_1	1 μm AMAD	5 μm AMAD	f_1	$e_{ing}(50)$
Bk-250	3.22h	M	5.0E-04	9.6E-10	7.1E-10	5.0E-04	1.4E-10
Californium							
Cf-244	0.323h	M	5.0E-04	1.3E-08	1.8E-08	5.0E-04	7.0E-11
Cf-246	1.49d	M	5.0E-04	4.2E-07	3.5E-07	5.0E-04	3.3E-09
Cf-248	334d	M	5.0E-04	8.2E-06	6.1E-06	5.0E-04	2.8E-08
Cf-249	3.50E+02y	M	5.0E-04	6.6E-05	4.5E-05	5.0E-04	3.5E-07
Cf-250	13.1y	M	5.0E-04	3.2E-05	2.2E-05	5.0E-04	1.6E-07
Cf-251	8.98E+02y	M	5.0E-04	6.7E-05	4.6E-05	5.0E-04	3.6E-07
Cf-252	2.64y	M	5.0E-04	1.8E-05	1.3E-05	5.0E-04	9.0E-08
Cf-253	17.8d	M	5.0E-04	1.2E-06	1.0E-06	5.0E-04	1.4E-09
Cf-254	60.5d	M	5.0E-04	3.7E-05	2.2E-05	5.0E-04	4.0E-07
Einsteinium							
Es-250	2.10h	M	5.0E-04	5.9E-10	4.2E-10	5.0E-04	2.1E-11
Es-251	1.38d	M	5.0E-04	2.0E-09	1.7E-09	5.0E-04	1.7E-10
Es-253	20.5d	M	5.0E-04	2.5E-06	2.1E-06	5.0E-04	6.1E-09
Es-254	276d	M	5.0E-04	8.0E-06	6.0E-06	5.0E-04	2.8E-08
Es-254m	1.64d	M	5.0E-04	4.4E-07	3.7E-07	5.0E-04	4.2E-09
Fermium							
Fm-252	22.7h	M	5.0E-04	3.0E-07	2.6E-07	5.0E-04	2.7E-09
Fm-253	3.00d	M	5.0E-04	3.7E-07	3.0E-07	5.0E-04	9.1E-10
Fm-254	3.24h	M	5.0E-04	5.6E-08	7.7E-08	5.0E-04	4.4E-10
Fm-255	20.1h	M	5.0E-04	2.5E-07	2.6E-07	5.0E-04	2.5E-09
Fm-257	101d	M	5.0E-04	6.6E-06	5.2E-06	5.0E-04	1.5E-08

Table B.1.—(*continued*)

		Effective dose coefficients (Sv Bq^{-1})					
		Inhalation, $e_{inh}(50)$				Ingestion	
Nuclide	$t_{1/2}$	Type	f_1	1 μm AMAD	5 μm AMAD	f_1	$e_{ing}(50)$
Mendelevium							
Md-257	5.20h	M	5.0E-04	2.3E-08	2.0E-08	5.0E-04	1.2E-10
Md-258	55.0d	M	5.0E-04	5.5E-06	4.4E-06	5.0E-04	1.3E-08

ANNEXE C. EFFECTIVE DOSE COEFFICIENTS FOR SOLUBLE OR REACTIVE GASES

Table C.1. Soluble or reactive gases (Class SR-1 and SR-2)

Nuclide/Chemical form	$t_{1/2}$	$e_{inh}(50)$ (Sv Bq^{-1})
Tritium gas [a]	12.3y	1.8E-15
Tritiated water [b]	12.3y	1.8E-11
Organically bound tritium	12.3y	4.1E-11
Carbon-11 vapour	0.34h	3.2E-12
Carbon-11 dioxide	0.34h	2.2E-12
Carbon-11 monoxide	0.34h	1.2E-12
Carbon-14 vapour	5.73E+03y	5.8E-10
Carbon-14 dioxide	5.73E+03y	6.5E-12
Carbon-14 monoxide	5.73E+03y	8.0E-13
Sulphur-35 vapour	87.4d	1.2E-10
Nickel-56 carbonyl	6.10d	1.2E-09
Nickel-57 carbonyl	1.50d	5.6E-10
Nickel-59 carbonyl	7.50E+04y	8.3E-10
Nickel-63 carbonyl	96.0y	2.0E-09
Nickel-65 carbonyl	2.52h	3.6E-10
Nickel-66 carbonyl	2.27d	1.6E-09
Iodine-120 vapour	1.35h	3.0E-10
Iodine-120m vapour	0.88h	1.8E-10
Iodine-121 vapour	2.12h	8.6E-11
Iodine-123 vapour	13.2h	2.1E-10
Iodine-124 vapour	4.18d	1.2E-08
Iodine-125 vapour	60.1d	1.4E-08
Iodine-126 vapour	13.0d	2.6E-08

[a] Irradiation from gas within the lungs might increase this value by about 20%.
[b] Dose from activity absorbed through the skin (as described in ICRP, 1979) is not included here.

Table C.1.—(*continued*)

Nuclide/Chemical form	$t_{1/2}$	$e_{inh}(50)$ (Sv Bq^{-1})
Iodine-128 vapour	0.42h	6.5E-11
Iodine-129 vapour	1.57E+07y	9.6E-08
Iodine-130 vapour	12.4h	1.9E-09
Iodine-131 vapour	8.04d	2.0E-08
Iodine-132 vapour	2.30h	3.1E-10
Iodine-132m vapour	1.39h	2.7E-10
Iodine-133 vapour	20.8h	4.0E-09
Iodine-134 vapour	0.88h	1.5E-10
Iodine-135 vapour	6.61h	9.2E-10
Mercury-193 vapour	3.50h	1.1E-09
Mercury-193m vapour	11.1h	3.1E-09
Mercury-194 vapour	2.60E+02y	4.0E-08
Mercury-195 vapour	9.90h	1.4E-09
Mercury-195m vapour	1.73d	8.2E-09
Mercury-197 vapour	2.67d	4.4E-09
Mercury-197m vapour	23.8h	5.8E-09
Mercury-199m vapour	0.71h	1.8E-10
Mercury-203 vapour	46.60d	7.0E-09

ANNEXE D. EFFECTIVE DOSE RATES FOR INERT GASES

Table D.1. Inert gases (Class SR-D)

Nuclide	$t_{1/2}$	Effective dose rate per unit air concentration ($Sv\ d^{-1}/Bq\ m^{-3}$)
Argon		
Ar-37	35.0 d	4.1E − 15
Ar-39	269 y	1.1E − 11
Ar-41	1.83 h	5.3E − 9
Krypton		
Kr-74	11.5 m	4.5E − 9
Kr-76	14.8 h	1.6E − 9
Kr-77	74.7 m	3.9E − 9
Kr-79	1.46 d	9.7E − 10
Kr-81	2.10E + 05 y	2.1E − 11
Kr-83m	1.83 h	2.1E − 13
Kr-85	10.7 y	2.2E − 11
Kr-85m	4.48 h	5.9E − 10
Kr-87	1.27 h	3.4E − 9
Kr-88	2.84 h	8.4E − 9
Xenon		
Xe-120	40.0 m	1.5E − 9
Xe-121	40.1 m	7.5E − 9
Xe-122	20.1 h	1.9E − 10
Xe-123	2.08 h	2.4E − 9
Xe-125	17.0 h	9.3E − 10
Xe-127	36.4 d	9.7E − 10
Xe-129m	8.0 d	8.1E − 11
Xe-131m	11.9 d	3.2E − 11
Xe-133m	2.19 d	1.1E − 10
Xe-133	5.24 d	1.2E − 10
Xe-135m	15.3 m	1.6E − 9
Xe-135	9.10 h	9.6E − 10
Xe-138	14.2 m	4.7E − 9

ANNEXE E. COMPOUNDS AND f_1 VALUES USED FOR THE CALCULATION OF INGESTION DOSE COEFFICIENTS

Table E.1.

Element	f_1	Compounds
Hydrogen	1.000	Ingestion of tritiated water
	1.000	Organically bound tritium
Beryllium	0.005	All compounds
Carbon	1.000	Labelled organic compounds
Fluorine	1.000	All compounds
Sodium	1.000	All compounds
Magnesium	0.500	All compounds
Aluminium	0.010	All compounds
Silicon	0.010	All compounds
Phosphorus	0.800	All compounds
Sulphur (inorganic)	0.800	Inorganic compounds
	0.100	Elemental sulphur
Sulphur (organic)	1.000	Sulphur in food
Chlorine	1.000	All compounds
Potassium	1.000	All compounds
Calcium	0.300	All compounds
Scandium	1.0×10^{-4}	All compounds
Titanium	0.010	All compounds
Vanadium	0.010	All compounds
Chromium	0.100	Hexavalent compounds
	0.010	Trivalent compounds
Manganese	0.100	All compounds
Iron	0.100	All compounds
Cobalt	0.100	Unspecified compounds
	0.050	Oxides, hydroxides and inorganic compounds
Nickel	0.050	All compounds
Copper	0.500	All compounds
Zinc	0.500	All compounds
Gallium	0.001	All compounds
Germanium	1.000	All compounds
Arsenic	0.500	All compounds
Selenium	0.800	Unspecified compounds
	0.050	Elemental selenium and selenides
Bromine	1.000	All compounds
Rubidium	1.000	All compounds
Strontium	0.300	Unspecified compounds
	0.010	Strontium titanate ($SrTiO_3$)
Yttrium	1.0×10^{-4}	All compounds
Zirconium	0.002	All compounds
Niobium	0.010	All compounds
Molybdenum	0.800	Unspecified compounds
	0.050	Molybdenum sulphide
Technetium	0.800	All compounds
Ruthenium	0.050	All compounds
Rhodium	0.050	All compounds
Palladium	0.005	All compounds
Silver	0.050	All compounds
Cadmium	0.050	All inorganic compounds
Indium	0.020	All compounds
Tin	0.020	All compounds
Antimony	0.100	All compounds

Table E.1.—(*continued*)

Element	f_1	Compounds
Tellurium	0.300	All compounds
Iodine	1.000	All compounds
Caesium	1.000	All compounds
Barium	0.100	All compounds
Lanthanum	5.0×10^{-4}	All compounds
Cerium	5.0×10^{-4}	All compounds
Praseodymium	5.0×10^{-4}	All compounds
Neodymium	5.0×10^{-4}	All compounds
Promethium	5.0×10^{-4}	All compounds
Samarium	5.0×10^{-4}	All compounds
Europium	5.0×10^{-4}	All compounds
Gadolinium	5.0×10^{-4}	All compounds
Terbium	5.0×10^{-4}	All compounds
Dysprosium	5.0×10^{-4}	All compounds
Holmium	5.0×10^{-4}	All compounds
Erbium	5.0×10^{-4}	All compounds
Thulium	5.0×10^{-4}	All compounds
Ytterbium	5.0×10^{-4}	All compounds
Lutetium	5.0×10^{-4}	All compounds
Hafnium	0.002	All compounds
Tantalum	0.001	All compounds
Tungsten	0.300	Unspecified compounds
	0.010	Tungstic acid
Rhenium	0.800	All compounds
Osmium	0.010	All compounds
Iridium	0.010	All compounds
Platinum	0.010	All compounds
Gold	0.100	All compounds
Mercury (inorganic)	0.020	All inorganic compounds
Mercury (organic)	1.000	Methyl mercury
	0.400	Unspecified organic compounds
Thallium	1.000	All compounds
Lead	0.200	All compounds
Bismuth	0.050	All compounds
Polonium	0.100	All compounds
Astatine	1.000	All compounds
Francium	1.000	All compounds
Radium	0.200	All compounds
Actinium	5.0×10^{-4}	All compounds
Thorium	5.0×10^{-4}	Unspecified compounds
	$2.0\ 10^{-4}$	Oxides and hydroxides
Protactinium	5.0×10^{-4}	All compounds
Uranium	0.020	Unspecified compounds
	0.002	Most tetravalent compounds, e.g. UO_2, U_3O_8, UF_4
Neptunium	5.0×10^{-4}	All compounds
Plutonium	5.0×10^{-4}	Unspecified compounds
	1.0×10^{-4}	Nitrates
	1.0×10^{-5}	Insoluble oxides
Americium	5.0×10^{-4}	All compounds
Curium	5.0×10^{-4}	All compounds
Berkelium	5.0×10^{-4}	All compounds
Californium	5.0×10^{-4}	All compounds
Einsteinium	5.0×10^{-4}	All compounds
Fermium	5.0×10^{-4}	All compounds
Mendelevium	5.0×10^{-4}	All compounds

ANNEXE F. COMPOUNDS, LUNG CLEARANCE TYPES AND f_1 VALUES USED FOR THE CALCULATION OF INHALATION DOSE COEFFICIENTS FOR WORKERS

Table F.1.

Element	Type	f_1	Compounds
Beryllium	M	0.005	Unspecified compounds
	S	0.005	Oxides, halides and nitrates
Fluorine	F	1.000	Determined by combining cation
	M	1.000	Determined by combining cation
	S	1.000	Determined by combining cation
Sodium	F	1.000	All compounds
Magnesium	F	0.500	Unspecified compounds
	M	0.500	Oxides, hydroxides, carbides, halides and nitrates
Aluminium	F	0.010	Unspecified compounds
	M	0.010	Oxides, hydroxides, carbides, halides, nitrates and metallic aluminium
Silicon	F	0.010	Unspecified compounds
	M	0.010	Oxides, hydroxides, carbides and nitrates
	S	0.010	Aluminosilicate glass aerosol
Phosphorus	F	0.800	Unspecified compounds
	M	0.800	Some phosphates: determined by combining cation
Sulphur	F	0.800	Sulphides and sulphates: determined by combining cation
	M	0.800	Elemental sulphur. Sulphides and sulphates: determined by combining cation
Chlorine	F	1.000	Determined by combining cation
	M	1.000	Determined by combining cation
Potassium	F	1.000	All compounds
Calcium	M	0.300	All compounds
Scandium	S	1.0×10^{-4}	All compounds
Titanium	F	0.010	Unspecified compounds
	M	0.010	Oxides, hydroxides, carbides, halides and nitrates
	S	0.010	Strontium titanate ($SrTiO_3$)
Vanadium	F	0.010	Unspecified compounds
	M	0.010	Oxides, hydroxides, carbides and halides
Chromium	F	0.100	Unspecified compounds
	M	0.100	Halides and nitrates
	S	0.100	Oxides and hydroxides
Manganese	F	0.100	Unspecified compounds
	M	0.100	Oxides, hydroxides, halides and nitrates
Iron	F	0.100	Unspecified compounds
	M	0.100	Oxides, hydroxides and halides
Cobalt	M	0.100	Unspecified compounds
	S	0.050	Oxides, hydroxides, halides and nitrates
Nickel	F	0.050	Unspecified compounds
	M	0.050	Oxides, hydroxides and carbides
Copper	F	0.500	Unspecified inorganic compounds
	M	0.500	Sulphides, halides and nitrates
	S	0.500	Oxides and hydroxides
Zinc	S	0.500	All compounds
Gallium	F	0.001	Unspecified compounds
	M	0.001	Oxides, hydroxides, carbides, halides and nitrates
Germanium	F	1.000	Unspecified compounds
	M	1.000	Oxides, sulphides and halides
Arsenic	M	0.500	All compounds
Selenium	F	0.800	Unspecified inorganic compounds
	M	0.800	Elemental selenium, oxides, hydroxides and carbides

Table F.1.—(*continued*)

Element	Type	f_1	Compounds
Bromine	F	1.000	Determined by combining cation
	M	1.000	Determined by combining cation
Rubidium	F	1.000	All compounds
Strontium	F	0.300	Unspecified compounds
	S	0.010	Strontium titanate ($SrTiO_3$)
Yttrium	M	1.0×10^{-4}	Unspecified compounds
	S	1.0×10^{-4}	Oxides and hydroxides
Zirconium	F	0.002	Unspecified compounds
	M	0.002	Oxides, hydroxides, halides and nitrates
	S	0.002	Zirconium carbide
Niobium	M	0.010	Unspecified compounds
	S	0.010	Oxides and hydroxides
Molybdenum	F	0.800	Unspecified compounds
	S	0.050	Molybdenum sulphide, oxides and hydroxides
Technetium	F	0.800	Unspecified compounds
	M	0.800	Oxides, hydroxides, halides and nitrates
Ruthenium	F	0.050	Unspecified compounds
	M	0.050	Halides
	S	0.050	Oxides and hydroxides
Rhodium	F	0.050	Unspecified compounds
	M	0.050	Halides
	S	0.050	Oxides and hydroxides
Palladium	F	0.005	Unspecified compounds
	M	0.005	Nitrates and halides
	S	0.005	Oxides and hydroxides
Silver	F	0.050	Unspecified compounds and metallic silver
	M	0.050	Nitrates and sulphides
	S	0.050	Oxides and hydroxides
Cadmium	F	0.050	Unspecified compounds
	M	0.050	Sulphides, halides and nitrates
	S	0.050	Oxides and hydroxides
Indium	F	0.020	Unspecified compounds
	M	0.020	Oxides, hydroxides, halides and nitrates
Tin	F	0.020	Unspecified compounds
	M	0.020	Stannic phosphate, sulphides, oxides, hydroxides, halides and nitrates
Antimony	F	0.100	Unspecified compounds
	M	0.010	Oxides, hydroxides, halides, sulphides, sulphates and nitrates
Tellurium	F	0.300	Unspecified compounds
	M	0.300	Oxides, hydroxides and nitrates
Iodine	F	1.000	All compounds
Caesium	F	1.000	All compounds
Barium	F	0.100	All compounds
Lanthanum	F	5.0×10^{-4}	Unspecified compounds
	M	5.0×10^{-4}	Oxides and hydroxides
Cerium	M	5.0×10^{-4}	Unspecified compounds
	S	5.0×10^{-4}	Oxides, hydroxides and fluorides
Praseodymium	M	5.0×10^{-4}	Unspecified compounds
	S	5.0×10^{-4}	Oxides, hydroxides, carbides and fluorides
Neodymium	M	5.0×10^{-4}	Unspecified compounds
	S	5.0×10^{-4}	Oxides, hydroxides, carbides and fluorides
Promethium	M	5.0×10^{-4}	Unspecified compounds
	S	5.0×10^{-4}	Oxides, hydroxides, carbides and fluorides
Samarium	M	5.0×10^{-4}	All compounds
Europium	M	5.0×10^{-4}	All compounds
Gadolinium	F	5.0×10^{-4}	Unspecified compounds
	M	5.0×10^{-4}	Oxides, hydroxides and fluorides
Terbium	M	5.0×10^{-4}	All compounds
Dysprosium	M	5.0×10^{-4}	All compounds
Holmium	M	5.0×10^{-4}	Unspecified compounds

Table F.1.—(*continued*)

Element	Type	f_1	Compounds
Erbium	M	5.0×10^{-4}	All compounds
Thulium	M	5.0×10^{-4}	All compounds
Ytterbium	M	5.0×10^{-4}	Unspecified compounds
	S	5.0×10^{-4}	Oxides, hydroxides and fluorides
Lutetium	M	5.0×10^{-4}	Unspecified compounds
	S	5.0×10^{-4}	Oxides, hydroxides and fluorides
Hafnium	F	0.002	Unspecified compounds
	M	0.002	Oxides, hydroxides, halides, carbides and nitrates
Tantalum	M	0.001	Unspecified compounds
	S	0.001	Elemental tantalum, oxides, hydroxides, halides, carbides, nitrates and nitrides
Tungsten	F	0.300	All compounds
Rhenium	F	0.800	Unspecified compounds
	M	0.800	Oxides, hydroxides, halides and nitrates
Osmium	F	0.010	Unspecified compounds
	M	0.010	Halides and nitrates
	S	0.010	Oxides and hydroxides
Iridium	F	0.010	Unspecified compounds
	M	0.010	Metallic iridium, halides and nitrates
	S	0.010	Oxides and hydroxides
Platinum	F	0.010	All compounds
Gold	F	0.100	Unspecified compounds
	M	0.100	Halides and nitrates
	S	0.100	Oxides and hydroxides
Mercury (inorganic)	F	0.020	Sulphates
	M	0.020	Oxides, hydroxides, halides, nitrates and sulphides
Mercury (organic)	F	0.400	All organic compounds
Thallium	F	1.000	All compounds
Lead	F	0.200	All compounds
Bismuth	F	0.050	Bismuth nitrate
	M	0.050	Unspecified compounds
Polonium	F	0.100	Unspecified compounds
	M	0.100	Oxides, hydroxides and nitrates
Astatine	F	1.000	Determined by combining cation
	M	1.000	Determined by combining cation
Francium	F	1.000	All compounds
Radium	M	0.200	All compounds
Actinium	F	5.0×10^{-4}	Unspecified compounds
	M	5.0×10^{-4}	Halides and nitrates
	S	5.0×10^{-4}	Oxides and hydroxides
Thorium	M	5.0×10^{-4}	Unspecified compounds
	S	2.0×10^{-4}	Oxides and hydroxides
Protactinium	M	5.0×10^{-4}	Unspecified compounds
	S	5.0×10^{-4}	Oxides and hydroxides
Uranium	F	0.020	Most hexavalent compounds, eg. UF_6, UO_2F_2 and $UO_2(NO_3)_2$
	M	0.020	Less soluble compounds, eg. UO_3, UF_4, UCl_4 and most other hexavalent compounds
	S	0.002	Highly insoluble compounds, e.g. UO_2 and U_3O_8
Neptunium	M	5.0×10^{-4}	All compounds
Plutonium	M	5.0×10^{-4}	Unspecified compounds
	S	1.0×10^{-5}	Insoluble oxides
Americium	M	5.0×10^{-4}	All compounds
Curium	M	5.0×10^{-4}	All compounds
Berkelium	M	5.0×10^{-4}	All compounds
Californium	M	5.0×10^{-4}	All compounds
Einsteinium	M	5.0×10^{-4}	All compounds
Fermium	M	5.0×10^{-4}	All compounds
Mendelevium	M	5.0×10^{-4}	All compounds

PREFACE

This summary of the basic principles for radiation protection of the patient in nuclear medicine was prepared by the Committee on Protection in Medicine of the International Commission on Radiological Protection (ICRP) to encourage medical professionals to become aware of and to utilise these basic principles. Nuclear medicine is that field of medical practice in which unsealed radioactive substances are used for diagnosis and therapy

A list of the references of the relevant publications is provided at the end of the summary. The references contain full discussions of the scientific and radiation protection considerations that contributed to the formulation of the basic principles. The reader is encouraged to consult the Commission's publications and the extensive references supporting these publications.

In 1977 the Commission published general recommendations on radiation protection in medicine in *ICRP Publication 26* (ICRP, 1977). In 1987 the Commission published detailed information for radiation protection of the patient in nuclear medicine in *ICRP Publication 52* (ICRP, 1987a). In 1987 and 1992 the Commission published detailed data on radiation dose to patients from radiopharmaceuticals in *ICRP Publication 53* (ICRP, 1987b) and *ICRP Publication 62* (ICRP, 1992). In 1991 the Commission published its new general recommendations, which included recommendations on radiation protection in medicine, in *ICRP Publication 60* (ICRP, 1991). This document summarises the information given in *ICRP Publication 52* and further updates the summary with the later information in *ICRP Publications 53*, *60* and *62*, particularly on the latest health effects and the updated system of radiological protection in *ICRP Publication 60*.

The membership of ICRP Committee 3 on Protection in Medicine during the preparation of this summary was:

J. Liniecki (Chairman)	S. Mattsson
C. F. Arias	M. Rosenstein (Secretary)
J. J. Conway	J. G. B. Russell
J. E. Gray	S. Somasundaram
M. Iio (deceased)	L. B. Sztanyik
J. Jankowski	J. Valentin
E. I. Komarov	G. A. M. Webb
A. Laugier	

The roles of Dr Conway and Dr Rosenstein are particularly acknowledged.

1. THE ICRP SYSTEM OF RADIOLOGICAL PROTECTION

In *ICRP Publication 60* (ICRP, 1991), the system of radiological protection recommended by the Commission for proposed and continuing practices is based on the following general principles:

(a) No practice involving exposures to radiation should be adopted unless it produces sufficient benefit to the exposed individuals or to society to offset the radiation detriment it causes. (The justification of a practice.)

(b) In relation to any particular source within a practice, the magnitude of individual doses, the number of people exposed, and the likelihood of incurring exposures where these are not certain to be received should all be kept as low as reasonably achievable, economic and social factors being taken into account. This procedure should be constrained by restrictions on the doses to individuals (dose constraints), so as to limit the inequity likely to result from the inherent economic and social judgments. (The optimisation of protection.)

(c) The exposure of individuals resulting from the combination of all the relevant practices should be subject to dose limits, or to some control of risk in the case of potential exposures.* These are aimed at ensuring that no individual is exposed to radiation risks that are judged to be unacceptable from these practices in any normal circumstances. Not all sources are susceptible to control by action at the source and it is necessary to specify the sources to be included as relevant before selecting a dose limit. (Individual dose and risk limits.)

Principles (a) (the justification of a practice) and (b) (the optimisation of protection) are relevant to radiation protection of the patient in medicine. The information in this summary identifies the application of these recommendations to nuclear medicine. Principle (c) (individual dose and risk limits) is not applicable to radiation doses received by patients from nuclear medicine: firstly, because they are direct recipients of benefits of the nuclear medicine procedure; and, secondly, because differing clinical problems in different nuclear medicine procedures must override any overall formula. If the procedure is justified and the protection is optimised, the dose in the patient will be compatible with the medical purposes.

2. HEALTH EFFECTS OF IONISING RADIATION

2.1. Absorbed Dose and Effective Dose

When ionising radiation interacts with matter, energy is absorbed, mainly by the process of ionisation. The mean energy imparted by ionising radiation per unit mass at a point in the human body is known as the absorbed dose in tissue. The unit of absorbed dose in the International System of Units is the gray (Gy), which is one joule per kilogram (J kg^{-1}). A commonly used subunit of the gray is the milligray (mGy), or 1/1000th of a gray.

Radiation energy absorbed in living tissues initiates physical and chemical reactions, resulting in biological changes. When the intent is to relate the absorbed dose to specific radiation-induced health effects, it is important to identify the particular body tissues in which

* In proposed and continuing practices, it will often be virtually certain that exposures will occur and their magnitude will be predictable, albeit with some uncertainty. Sometimes, however, there will be a potential for exposure, but no certainty that it will occur. The Commission calls such exposures "potential exposures".

absorbed doses occur. In this summary, when specific tissues are discussed, the term "absorbed dose" is used and the relevant tissue or location is identified; the term "dose" is used when the intended meaning is more general.

When parts of the body are irradiated during nuclear medicine procedures individual tissues and organs are likely to receive different absorbed doses. In order to facilitate a comparison between doses from different types of nuclear medicine procedures, some quantity which reflects the modifying effects of different types of radiation and the relative radiosensitivity of the irradiated tissues and organs is required. This quantity is defined as the absorbed dose weighted by a radiation weighting factor which is selected for the type and energy of the radiation incident upon the body, or in the case of sources within the body, emitted by the source; which is further weighted by a tissue weighting factor, which represents the relative contribution of the tissue or organ detriment to the total detriment, as if the whole body were uniformly irradiated. This doubly weighted absorbed dose is called the effective dose. The unit of effective dose is the sievert (Sv) which is 1 J kg^{-1}. A commonly used subunit of the sievert is the millisievert (mSv), or 1/1000th of a sievert.

2.2. Risk of Deterministic Effects

Most organs and tissues of the body are unaffected by the loss of even substantial numbers of cells, but if the number lost is large enough, there will be observable harm reflecting a loss of tissue function. The probability of causing such harm will be zero at small doses, but above some level of dose (the threshold) will increase steeply to unity (100%). Above the threshold, the severity of the harm will also increase with dose. This type of effect, previously called "non-stochastic", is now called "deterministic" by the Commission.

Some nuclear medicine procedures, particularly at the absorbed doses delivered in therapy, can produce localised cellular reactions, which are deterministic in character. However, in properly conducted diagnostic nuclear medicine procedures, these deterministic effects do not occur because the absorbed doses are well below the threshold for such effects.

2.3. Risk of Neoplasia and Hereditary Effects

In contrast, there may be no threshold of absorbed dose for the initiation of some deleterious biological changes. Consequently, even small absorbed doses in tissues may increase the risk of cancer and small absorbed doses in the gonads may induce mutations or chromosomal changes, potentially capable of inducing hereditary disorders in the progeny of irradiated parents. These types of effect are known as stochastic; i.e. the probability of occurrence of the effect increases with the absorbed dose, whereas the severity of the effect is independent of the dose. It therefore is assumed in radiation protection planning that a threshold of absorbed dose for stochastic effects is unlikely and every increment of absorbed dose to an individual may carry some risk.

The nominal risks (i.e. the probability of occurrence) for radiation-induced serious hereditary effects and fatal cancers for the general population, discussed in *ICRP Publication* 60, are given in Table 1. In addition, non-fatal cancers occur. Therefore, nominal (typical) ratios for total cancers (fatal plus non-fatal) to fatal cancers for each type are also given in Table 1. The nominal risks for infants and children and for adults at certain ages are different than for the general population. Footnotes c and d to Table 1 provide an estimate of the magnitude of these differences for severe hereditary disorders and the sum of fatal cancers for whole-body irradiation. It should be recognised from Table 1 that the baseline cancer mortality varies from country to country.

Table 1. Nominal risks[a] at low doses and low dose rates for low-LET[b] radiation expressed as severe hereditary disorders and fatal cancers (lifetime)

Effect	Nominal risk per milligray	Nominal ratio: total cancers to fatal cancers
Hereditary (gonads)[c]	10×10^{-6} (all generations)	—
Cancers	*Fatality probability*	
Leukaemia (active bone marrow)	5×10^{-6}	1.01
Bone surface	0.5×10^{-6}	1.4
Breast (females only)	4×10^{-6}	2.0
Lung	8.5×10^{-6}	1.05
Thyroid	0.8×10^{-6}	10
Colon	8.5×10^{-6}	1.8
Oesophagus	3×10^{-6}	1.05
Skin	0.2×10^{-6}	500
Stomach	11×10^{-6}	1.1
Liver	1.5×10^{-6}	1.05
Bladder	3×10^{-6}	2.0
Ovary (females only)	2×10^{-6}	1.4
Other (combined remaining tissues and organs)	5×10^{-6}	1.8
Sum of fatal cancers for whole-body irradiation (average for male and female)[d]	50×10^{-6} (1 in 20,000)	
Baseline cancer mortality	0.15 (1 in 6.7) to 0.25 (1 in 4)	

[a]The nominal risks are average values for a population comprised of equal numbers of males and females of all ages (except for the breast and ovary, which are for females only).

[b]LET means linear energy transfer; low-LET radiation refers to sparsely ionising radiations such as gamma rays, x rays and beta particles.

[c]For infants and children with a full expectation of reproductive potential, the nominal risk per milligray for severe hereditary disorders is a few times larger than the value given. For older age groups, the nominal risk milligray would decrease towards zero as the expectation of reproductive potential decreases.

[d]For infants and children who are irradiated, the nominal risk is very likely to be 2 to 3 times higher than the value of 50×10^{-6} per millgray given for whole-body irradiation and the risk appears after a long latency (except for some types of leukaemia). For adults aged over 50 years at the time of irradiation, the nominal risk is 5 to 10 times lower than the value of 50×10^{-6} per milligray given for whole-body irradiation. This lower risk reflects the latency associated with radiation-induced cancers other than some types of leukaemia.

2.4. Effects Following Irradiation *In Utero*

The potential for effects on the conceptus of irradiation depends on the time of the irradiation relative to conception. When the number of cells in the conceptus is small and their nature is not yet specialised, the effect of damage to these cells is most likely to take the form of a failure to implant or of an undetectable death of the conceptus. It is thought that any cellular damage at this stage is much more likely to cause the death of the conceptus than to result in stochastic effects expressed in the live-born. Irradiation of the embryo in the first three weeks following conception is not likely to result in deterministic or stochastic effects in the live-born child, despite the fact that the central nervous system and the heart are beginning to develop in the third week. During the rest of the period of major organogenesis, malformations may be caused in the organ under development at the time of irradiation. These effects are deterministic in character with a threshold in man, estimated from animal experiments, to be about 0.1 Gy. Two specific effects of radiation on the developing embryo or fetus need consideration, namely severe mental retardation and cancers which may be expressed during childhood or in adult life. The periods of sensitivity after conception for the types of effects mentioned in this paragraph which are discussed in *ICRP Publication* 60, are summarised in Table 2.

Table 2. Types of effects following irradiation *in utero*

Time after conception	Effect	Normal incidence in live-born
First three weeks	No deterministic or stochastic effects in live-born child	–
3rd through 8th weeks	Potential for malformation of organs[a]	0.06 (1 in 17)
8th through 25th weeks	Potential for severe mental retardation[b]	5×10^{-3} (1 in 200)
4th week throughout pregnancy	Cancer in childhood or in adult life[c]	1×10^{-3} (1 in 1000)

[a]Malformation of organs appears to be a deterministic effect, with a threshold dose in man, estimated from animal experiments to be about 0.1 Gy.

[b]The risk for severe mental retardation is associated by the Commission with an observed shift in intelligence quotient (IQ) (i.e. 30 IQ units per 1 gray in the brain during the period 8th through 15th week after conception with lesser IQ shifts during the period 16th through 25th week). At absorbed doses in the brain of the order of 0.1 gray, no effect would be detectable in the general distribution of IQ in an irradiated group or population, but at somewhat larger absorbed doses, the effect might be sufficient to show an increase in the number of children classified as severely retarded.

[c]The risk of fatal cancers expressed in childhood or in adult life for individuals irradiated *in utero* may be similar to the risk to individuals irradiated in the first decade of life, which is very likely somewhat larger than that for the population as a whole (see entry for sum of fatal cancers for whole-body irradiation in Table 1 and footnote d).

3. RADIATION DOSE FROM NUCLEAR MEDICINE

The principal sources of absorbed dose for members of the public are natural radiation and medical applications. The contribution from all medical uses to the annual *per capita* dose varies from a few percent of the dose from natural background in developing countries to substantially higher percentages in developed countries. The largest part of this contribution comes from diagnostic radiology. A relatively small contribution comes from nuclear medicine.

3.1. Absorbed Dose

Radiations emitted by radiopharmaceuticals are absorbed by organs and tissues of the body, and the magnitude of the resulting absorbed doses depends on several factors. These include the amount of radioactive material administered, the biokinetics and the decay scheme of the radionuclide.

In general, diagnostic applications of nuclear medicine procedures result in absorbed doses to individual organs not exceeding a few tens of milligrays. Therepeutic applications, aimed at total or partial destruction of a tissue, deliver much higher absorbed doses in target organs, of the order of tens to hundreds of grays. Therefore, whereas misapplication of a radiopharmaceutical at diagnostic levels carries a relatively small risk of tissue damage to the patient, inappropriate application or dosimetric error in therapy may result in unacceptable tissue damage or inadequate treatment.

The biokinetics of an intravenously administered radiopharmaceutical can sometimes be modified by actions taken before or after injection. Thus, means taken to reduce the absorbed dose, even after administration of the radiopharmaceutical, can be useful and should be undertaken when possible.

Absorbed doses delivered to the embryo or fetus from radionuclides contained in, or passing through, the neighbouring maternal organs can be higher than the average to the mother's body. Some radiopharmaceuticals may pass through the placenta to the fetal circulation. The nuclear medicine physician must therefore proceed with special caution when administering radiopharmaceuticals to pregnant women.

3.2. Administered Activity

Estimation of the optimum activity of a radiopharmaceutical for a given diagnostic procedure (especially when applied to an individual patient) is a complex matter. The amount will depend on the type of equipment used, the body build and weight of the individual patient, the patient's metabolic characteristics and clinical condition, and sometimes on the experience of the individual physician who will evaluate the image.

For a given type of imaging equipment the diagnostic value of the information obtained from a test involving the use of a radiopharmaceutical will vary with the amount of administered activity. There is a threshold of the administered activity below which no useful information can be expected. Above this level, the diagnostic quality (and therefore usefulness of the image) will increase steeply with the increasing activity. Once an acceptable quality of the image has been reached, further increase of the administered activity will only increase the absorbed dose and not the value of the diagnostic information.

The administration of amounts of activity that are substantially larger than the optimum, in order to improve marginally the quality of the results obtained, should be discouraged.

Limiting the administered amount of activity below the optimum, even for well-intentioned reasons, will usually lead to poor quality of the image, which may cause serious diagnostic errors.

Of equal importance to the choice of the optimum amount of administered activity is the avoidance of failure to obtain the diagnostic information; failure results in an unnecessary irradiation and may also necessitate repetition of the test.

3.3. Estimates of Absorbed Dose

Estimates of the absorbed dose in organs and tissues depend upon information on radionuclide decay, the amount of administered activity and the distribution and retention of the radionuclide in the body. Biokinetic models and data for approximately 150 substances are compiled in *ICRP Publication 53*.

The biological data available are used to calculate the activity in the various source organs included in the biokinetic model. This enables the calculation of the total absorbed dose in each selected target organ from the radioactive material in these source organs. This formalism is described in detail in *ICRP Publication 53*.

The clinician who uses these or other absorbed dose data as a guide needs to realise that the physical factors used in the calculations are based upon a mathematical model of a Reference Man and not on the individual patient being considered. Published biokinetic data for humans are sparse. Thus, the biological data used to derive the absorbed dose may lead only to an approximate estimate in a given clinical case.

The data for absorbed doses in organs are usually presented as mean values over the whole organ. No attempt has been made to take gross or subcellular non-uniformities into account in the absorbed doses presented.

Table 3 contains absorbed doses for a selection of clinically used radiopharmaceuticals administered to adults. Some of the radiopharmaceuticals listed are not currently in use in most developed countries, but are shown because they are still being used in many parts of the world. The absorbed doses in milligray (mGy) are expressed per megabecquerel (MBq, i.e. 10^6 becquerel) of administered activity, and, therefore, the actual absorbed doses received will depend on the total activity administered. Absorbed doses are shown for the tissues or organs receiving the highest absorbed doses; in every case the absorbed dose to the uterus is shown, as an indication of the absorbed dose to a possible embryo or fetus (referred to as the conceptus). This information is from *ICRP Publication 53*, except for the information for ^{99m}Tc HMPAO and $^{99m}TcMAG_3$, which is from *ICRP Publication 62*.

When radiopharmaceuticals are administered, individual organs may receive very different absorbed doses. In order to facilitate a comparison between different types of nuclear medicine procedures, the doses can be expressed in terms of effective dose (see Section 2.1). Effective doses in millisievert (mSv) are provided in Table 3, and are also expressed per megabecquerel (MBq) of administered activity. The information is from *ICRP Publication 62* and was computed using the formulation of effective dose in *ICRP Publication 60*.

The effective dose is a measure of the radiation detriment to the general population, averaged over the full age distribution of the general population and equal numbers of both sexes. Therefore, the effective dose can only be an approximate indicator of the relative radiation detriment to patients of specific sex and/or age groups from various nuclear medicine procedures.

4. NUCLEAR MEDICINE IN DIAGNOSIS

4.1. Distribution of Responsibilities

4.1.1. *Referring physician*

The referring physician should describe the patient's condition and indicate the clinical objectives, so that the nuclear medicine physician and the nuclear medicine physician's staff can carry out whichever procedure they judge is in the best interest of the patient. The referring physician's primary and proper concern is with the efficacy of the nuclear medicine test, i.e. will it contribute essentially to the diagnosis.

4.1.2. *Nuclear medicine physician*

The nuclear medicine physician has the ultimate responsibility for the control of all aspects of the conduct and extent of nuclear medicine examinations, including the protection of the patient from unnecessary radiation. The nuclear medicine physician should advise on the appropriateness of examinations, and determine the techniques to be used. Ideally, all nuclear medicine examinations should be individually planned so that the necessary information is obtained with minimum irradiation.

It is the responsibility of the nuclear medicine physician to assure that the necessary information is available to determine whether or not the procedure is justified.

Furthermore, a request for *in vivo* nuclear medicine procedures should first take account of the availability, relative efficacy and associated risk of alternative methods. These include techniques not utilising ionising radiations (e.g. ultrasound and magnetic resonance imaging), as well as those utilising ionising radiations (i.e. radiography and x-ray computed tomography).

4.1.3. *Medical physicist*

The medical physicist is responsible for the physical and technical aspects of radiation dosimetry, nuclear medicine instrumentation and radiation protection, and for quality control as well as for data-handling and computations arising from nuclear medicine tests.

4.1.4. *Technologist and other staff*

The main duty of the nuclear medicine technologist and other appropriately qualified members of staff is to carry out nuclear medicine examinations under the supervision of an appropriately trained physician.

4.1.5. *Committee on radiation protection*

The organisation of nuclear medicine services would benefit from some form of committee whose responsibilities include providing advice on the radiation protection of patients. The requirement for such a committee depends on the existing local, regional or national practice. The membership of the committee should include those knowledgeable about the effects of ionising radiation, the appropriate methods of protection, and social and legal matters pertaining to radiation. The committee should operate under a written policy and maintain appropriate records.

4.2. General Nuclear Medicine Procedures

4.2.1. *Choice of radiopharmaceutical*

If more than one radiopharmaceutical can be used for a procedure, consideration should be given to the physical, chemical and biological properties of each radiopharmaceutical, so as to minimise the absorbed dose and other risks to the patient, while at the same time providing the desired diagnostic information. Availability, shelf-life, instrumentation, and comparative costs are other factors that affect the choice.

The paediatric patient should be of special concern because of physical and biological factors that are unique to the child. The uptake, biologic distribution and retention of radiopharmaceuticals vary with age, the differences from the average values being most pronounced in childhood, especially in the neonate and infant. Furthermore, the risk resulting from a given absorbed dose to a child is likely to be higher than to an adult.

4.2.2. *Monitoring of the procedure*

While a study is being performed, the images should be reviewed by the responsible nuclear medicine physician, to ensure that the images are of acceptable quality and that the necessary information is being acquired.

4.2.3. *Immobilisation of the patient*

While most adults can maintain a required position without restraint or sedation during

Table 3. Absorbed dose (mGy) and effective dose (mSv) per unit administered activity (MBq) in normal adults

Function or organ examined	Radio-nuclide	Pharmaceutical	Most highly irradiated organs: Organ 1 (mGY/MBq)	Organ 2 (mGY/MBq)	Organ 3 (mGY/MBq)	Conceptus (mGY/MBq)	Effective dose (mSv/MBq)
Bone	^{99m}Tc	phosphate/ phosphonate	6.3×10^{-2} (bone surface)	5.0×10^{-2} (bladder)	9.6×10^{-3} (red marrow)	6.1×10^{-3}	5.8×10^{-3}
Renal (normal renal function)	^{51}Cr	EDTA	2.3×10^{-2} (bladder)	2.8×10^{-3} (uterus)	1.8×10^{-3} (kidneys)	2.8×10^{-3}	2.1×10^{-3}
	^{123}I	hippurate	2.0×10^{-1} (bladder	1.7×10^{-2} (uterus)	7.3×10^{-3} (lower large intestine)	1.7×10^{-2}	1.2×10^{-2}
	^{131}I	hippurate	9.6×10^{-1} (bladder)	3.5×10^{-2} (uterus)	3.0×10^{-2} (kidneys)	3.5×10^{-2}	5.3×10^{-2}
	^{99m}Tc	DTPA	6.5×10^{-2} (bladder)	7.9×10^{-3} (uterus)	4.4×10^{-3} (kidneys)	7.9×10^{-3}	5.2×10^{-3}
	^{99m}Tc	DMSA	1.7×10^{-1} (kidneys)	1.9×10^{-2} (bladder)	1.3×10^{-2} (adrenals, spleen)	4.6×10^{-3}	8.7×10^{-3}
	^{99m}Tc	MAG_3	1.1×10^{-1} (bladder)	1.2×10^{-2} (uterus)	5.7×10^{-3} (lower large intestine)	1.2×10^{-2}	7.3×10^{-3}
Thyroid	^{99m}Tc	pertechnetate (no blocking)	6.2×10^{-2} (upper large intestine	2.9×10^{-2} (stomach)	2.3×10^{-2} (thyroid)	8.1×10^{-3}	1.2×10^{-2}
	^{131}I	iodide (35% uptake)	5.0×10^{-2} (thyroid)	4.6×10^{-1} (stomach)	4.0×10^{-1} (bladder)	5.0×10^{-2}	24
	^{123}I	iodide (35% uptake)	4.5 (thyroid)	6.8×10^{-2} (stomach)	6.0×10^{-2} (bladder)	1.4×10^{-2}	2.2×10^{-1}
Liver (+gall bladder)	^{99m}Tc	colloid (large)	7.7×10^{-2} (spleen)	7.4×10^{-2} (liver)	1.2×10^{-2} (pancreas)	1.9×10^{-3}	9.2×10^{-3}
	^{99m}Tc	colloid (small)	7.7×10^{-2} (spleen)	7.4×10^{-2} (liver)	1.5×10^{-2} (red marrow)	1.8×10^{-3}	9.7×10^{-3}
	^{99m}Tc	HIDA	1.1×10^{-1} (gall bladder)	9.2×10^{-2} (upper large intestine)	6.2×10^{-2} (lower large intestine)	1.3×10^{-2}	1.5×10^{-2}
	^{57}Co	B_{12} (no carrier)	5.1×10 (liver)	5.4 (adrenals, pancreas)	5.0 (kidneys)	1.8	4.4
Brain	^{99m}Tc	pertechnetate (blocked thyroid)	3.2×10^{-2} (bladder)	6.6×10^{-3} (uterus)	4.7×10^{-3} (kidneys, ovaries)	6.6×10^{-3}	4.7×10^{-3}
	^{99m}Tc	gluconate/ glucoheptonate	5.6×10^{-2} (bladder)	4.9×10^{-2} (kidneys)	7.7×10^{-3} (uterus)	7.7×10^{-3}	5.4×10^{-3}
	^{99m}Tc	HMPAO	3.4×10^{-2} (kidneys)	2.6×10^{-2} (thyroid)	2.3×10^{-2} (bladder)	6.6×10^{-3}	9.3×10^{-3}
	^{18}F	FDG	1.7×10^{-1} (bladder)	6.5×10^{-2} (heart)	2.6×10^{-2} (brain)	2.0×10^{-2}	2.0×10^{-2}

Table 3 (*continued*)

Function or organ examined	Radio-nuclide	Pharmaceutical	Most highly irradiated organs: Organ 1 (mGY/MBq)	Organ 2 (mGY/MBq)	Organ 3 (mGY/MBq)	Conceptus (mGY/MBq)	Effective dose (mSv/MBq)
Lung	^{99m}Tc	MAA	6.7×10^{-2} (lungs)	1.6×10^{-2} (liver)	1.0×10^{-2} (bladder)	2.4×10^{-3}	1.1×10^{-2}
	^{99m}Tc	aerosol (fast clearance)	4.7×10^{-2} (bladder)	1.7×10^{-2} (lungs)	5.9×10^{-3} (uterus)	5.9×10^{-3}	6.1×10^{-3}
	^{99m}Tc	aerosol (slow clearance)	9.3×10^{-2} (lungs)	1.3×10^{-2} (bladder)	6.4×10^{-3} (breasts)	1.7×10^{-3}	1.4×10^{-2}
	^{133}Xe	gas, 5 min (re-breathing)	1.1×10^{-3} (lungs)	8.4×10^{-4} (red marrow)	8.3×10^{-4} (breasts)	7.4×10^{-4}	7.3×10^{-4}
	^{133}Xe	gas, 30 s 1 breath	7.7×10^{-4} (lungs)	1.2×10^{-4} (bone surfaces, red marrow, breasts)	1.1×10^{-4} (small and large intestines, liver, pancreas, spleen, uterus)	1.1×10^{-4}	1.8×10^{-4}
	^{81m}Kr	gas	2.1×10^{-4} (lungs)	4.6×10^{-6} (breasts)	3.5×10^{-6} (pancreas)	1.3×10^{-7}	9.5×10^{-4}
Heart	^{201}Tl	thallous ion	5.6×10^{-1} (testes)	5.4×10^{-1} (kidneys)	3.6×10^{-1} (lower large intestine)	5.0×10^{-2}	2.3×10^{-1}
	^{99m}Tc	RBC	2.3×10^{-2} (heart)	1.5×10^{-2} (sleen)	1.4×10^{-2} (lungs)	4.7×10^{-3}	6.6×10^{-3}
Abscess	^{111}In	white cells	5.5 (spleen)	7.1×10^{-1} (liver)	6.9×10^{-1} (red marrow)	1.2×10^{-1}	3.6×10^{-1}
	^{67}Ga	citrate	5.9×10^{-1} (bone surfaces)	2.0×10^{-1} (lower large intestine)	1.9×10^{-1} (red marrow)	7.9×10^{-2}	1.1×10^{-1}
Thrombi	^{125}I	fibrinogen (thyroid totally blocked)	3.2×10^{-1} (heart)	2.4×10^{-1} (spleen)	2.3×10^{-1} (lungs)	5.5×10^{-2}	8.0×10^{-2}
	^{111}In	platelets	7.5 (spleen)	7.3×10^{-1} (liver)	6.6×10^{-1} (pancreas)	9.5×10^{-2}	3.9×10^{-1}
	^{51}Cr	platelets	2.6 (spleen)	3.0×10^{-1} (liver)	1.9×10^{-1} (red marrow)	2.8×10^{-2}	1.4×10^{-1}
Pancreas	^{75}Se	methionine	6.2 (liver)	5.3 (kidneys)	3.9 (spleen)	2.6	2.5
Adrenals	^{75}Se	methyl cholesterol	5.1 (adrenals)	2.0 (liver)	1.8 (small intestine, pancreas, uterus, red marrow)	1.8	1.5
	^{131}I	MIBG	8.3×10^{-1} (liver)	5.9×10^{-1} (bladder)	4.9×10^{-1} (spleen)	8.0×10^{-2}	1.4×10^{-1}
Spleen	^{51}Cr	RBC denatured	5.6 (spleen)	3.0×10^{-1} (pancreas)	1.7×10^{-1} (liver)	1.3×10^{-2}	1.8×10^{-1}
GIT	^{99m}Tc	pertechnetate (oral, no blocking agent)	7.4×10^{-2} (upper large intestine)	5.0×10^{-2} (stomach)	3.0×10^{-2} (small intestine)	8.7×10^{-3}	1.4×10^{-2}

nuclear medicine procedures, it may be necessary to immobilise or sedate children so that the examination can be completed successfully.

4.2.4. *Methods for reducing absorbed dose to the patient*

Substantial reduction in absorbed dose from radiopharmaceuticals can often be readily achieved by the following simple measures:

— Oral or intravenous hydration of the patient is an effective means to reduce the absorbed dose from radiopharmaceuticals that are excreted by the kidneys. The patient should be encouraged to void frequently, especially in the immediate interval following the examination.

— The use of thyroid blocking agents such as KI or $KClO_4$ is a generally accepted practice when using radioactive iodine compounds or ^{99m}Tc as pertechnetate (except when used for thyroid imaging). Such agents can be administered after the data have been acquired (e.g. in Meckel's diverticulum scintigraphy) and still achieve a reduction in absorbed dose to the thyroid gland.

— Laxatives can be used to increase the elimination rate of radiopharmaceuticals or of their metabolites introduced or secreted into the gastrointestinal tract.

— The storage or retention of radiopharmaceuticals within specific organs can be influenced by drugs such as diuretics or gall bladder stimulants. Such interventions should be introduced into clinical protocols whenever they do not interfere with the acquisition of necessary clinical information, and are not hazardous to the patient.

A number of examples of methods appropriate for specific radiopharmaceuticals are listed in Table 4.

4.3. Nuclear Medicine Procedures on Women

4.3.1. *Women who may be pregnant*

In women of child-bearing age, the possibility of pregnancy and the justification for the examination should be considered. The recommended precautions to prevent or minimise irradiation of an embryo or fetus include the following:

— The patient must be carefully interviewed to assess the likelihood of pregnancy. Particular discretion is required to ascertain the possibility of pregnancy in an adolescent.

— It is prudent to consider as pregnant any woman of reproductive age presenting for a nuclear medicine examination at a time when a menstrual period is overdue or missed, unless there is information that precludes pregnancy (e.g. hysterectomy). If the menstrual cycle is irregular, a pregnancy test may be indicated before proceeding.

— In order to minimise the frequency of unintentional irradiation of the embryo or fetus, advisory notices should be posted at several places within the nuclear medicine department, and particularly at its reception area. For example:

IF IT IS POSSIBLE THAT YOU MIGHT BE PREGNANT,
NOTIFY STAFF BEFORE BEGINNING YOUR STUDY OR TREATMENT

4.3.2. *Pregnant women*

Irradiation of the fetus results from placental transfer and distribution of radiopharma-

Table 4. Methods used for the reduction of organ doses in various nuclear medicine procedures

Method	Radioisotope	Procedure	Organ
Hydration and frequent voiding	^{99m}Tc-phosphates	Bone scintigraphy Myocardial scintigraphy	Bladder region
	^{99m}Tc-DTPA	Renal scintigraphy	Bladder region
	^{131}I-hippurate	Renography	Bladder region
	^{99m}Tc-O_4^-	Radionuclide angiography	Bladder region
	^{111}In-DTPA	Cisternography	Bladder region
	^{99m}Tc-O_4^-	Scrotal scintigraphy	Bladder region
	^{99m}Tc-microspheres	Lung perfusion scintigraphy	Bladder region
	^{201}Tl-chloride	Cardiac scintigraphy	Bladder region
$KclO_4$ given before administration of radiopharmaceutical	^{99m}Tc-O_4^-	Radionuclide angiography Dacrocystography	Thyroid
$KClO_4$ administration after data acquisition	^{99m}Tc-O_4^-	Meckel's diverticulum scintigraphy Salivary gland scintigraphy	Throid and salivary glands
KI or $KClO_4$	^{131}I MIBG	Scintigraphy of adrenal medulla	Thyroid
	^{131}I Rose Bengal	Hepatobiliary scintigraphy	Thyroid
Catheterisation of infant's neurogenic bladder	^{99m}Tc-O_4^-	Radionuclide cystography	Bladder
Laxatives	^{67}Ga citrate	Scintigraphic detection of neoplastic and inflammatory foci	Bowel
Diuretics	^{99m}Tc-DTPA	Renal scintigraphy	Kidneys
	^{131}I-hippurate	Renography	Kidneys
	^{99m}Tc-MAG_3	Renography	Kidneys
Cholecystokinin (fatty meal)	^{99m}Tc-IDA compounds	Hepatobiliary scintigraphy	Gall bladder

ceuticals in the fetal tissues, or from external irradiation from the mother's organs and tissues. The chemical and biological properties of the radiopharmaceutical are the critical factors in possible placental transfer.

Some radiopharmaceuticals cross the placenta freely, e.g. radioactive iodides, and are taken up in fetal tissues, where they irradiate the tissues. Some analogues of natural metabolites (e.g. radiostrontium for calcium, and radiocaesium for potassium) are less readily transferred. Radiocolloids that are retained by the reticuloendothelial system of the mother, and do not cross the placenta, act as external sources of irradiation to the fetus.

In the case of radiopharmaceuticals that are rapidly eliminated by the kidneys, the urinary bladder, acting as a reservoir, can become a major source of irradiation to other organs and tissues and to the fetus. After the administration of radiopharmaceuticals, frequent voiding should, therefore, be ensured. This contribution to the fetal dose can be further reduced by administering the radiopharmaceutical when the bladder is partially filled, rather than immediately after voiding.

When a nuclear medicine examination is proposed for a pregnant woman, great care has to be taken to ascertain that the examination is indeed indicated. If, in consultation with the referring physician, it is deemed that the risk of not making a necessary diagnosis is greater than that of irradiating the fetus, the examination should be performed. The adsorbed dose to the fetus can be reduced by using smaller administered activities and longer imaging times. When ultrasound diagnostics are available, radionuclide studies for localisation of the placenta should be discouraged.

Irradiation of the pregnant patient, at a time when the pregnancy was unrecognised, often leads to her apprehension, because of concern about possible effects on the fetus. Even though the absorbed doses to the conceptus are generally small, such concern may lead to a suggestion that the pregnancy be terminated. However, on the basis of the relative risk increment, fetal irradiation from a usual diagnostic procedure does not justify terminating a pregnancy. When such a concern arises, an estimate of the absorbed dose, and the associated risk to the fetus, should be made by a qualified expert. With such expert and carefuly worded advice, the patient should then be in a position to make her own decision regarding continuation of pregnancy.

4.3.3. *Women who are breast feeding*

Since many radiopharmaceuticals are secreted in breast milk, it is safest to assume that, unless there are data to the contrary, some radioactive compounds will be found in the breast milk when a radiopharmaceutical is administered to a lactating female. Consideration should be given to postponing the procedure. If the procedure is performed, the child should not be breast-fed until the radiopharmaceutical is no longer secreted in an amount estimated to give an unacceptable absorbed dose to the child. In order to minimise the irradiation of the breat-fed child, advisory notices should be posted within the nuclear medicine department. An example of such a notice is:

PLEASE NOTIFY THE STAFF
IF YOU ARE BREAST FEEDING YOUR CHILD

It is recommended that the following actions should then be taken for various radiopharmaceuticals:

Group I: Stop nursing for at least 3 weeks

— All ^{131}I- and ^{125}I-radiopharmaceuticals except labelled hippurate
— ^{22}Na, ^{67}Ga, ^{201}Tl

Group II; Stop nursing for at least 12 hours

— ^{131}I-, ^{125}I- and ^{123}I-hippurate
— All ^{99m}Tc-compounds except labelled red blood cells, -phosphonates and -DTPA

Group III: Stop nursing for at least 4 hours

— ^{99m}Tc-red blood cells, -phosphonate and -DTPA

Group IV: No necessity to stop nursing

— ^{51}Cr-EDTA

4.3.4. *Avoidance of pregnancy after a diagnostic procedure*

Patients sometimes inquire about the time that should elapse between the completion of a diagnostic nuclear medicine procedure and attempts to become pregnant. Since there are no currently used diagnostic tests in which tissue radionuclide concentrations have sufficiently long effective half lives to irradiate a subsequent embryo significantly, there is no medical reason to wait.

4.4. Protection of the Family

Due to the short effective half-life of most diagnostic radiopharmaceuticals and their sporadic use, there is usually very little radiation hazard to the patient's family. Table 5 shows absorbed dose rates at various times and at various distances from individuals who have received diagnostic levels of various radiopharmaceuticals.

Table 5. Illustration of absorbed dose rates at various distances and times from a typical adult patient after administration of a radiopharmaceutical

			Absorbed dose rate (nGy/hr per MBq)**					
		Typical range of administered	Immediately after			After 2 hr		
Study	Radiopharmaceutical	activity (Mbq)	close*	0.3 m	1 m	close*	0.3 m	1 m
Bone scintigraphy	^{99m}Tc MDP	150–600	27	13	4	13	7	2
Liver scintigraphy	^{99m}Tc colloid	10–250	27	13	4	20	10	3
Blood pool determination	^{99m}Tc RBC	550–740	27	13	4	20	10	3
Myocardial scintigraphy	^{201}Tl	50–110	36	18	6	36	18	6

* At the body surface over the relevant tissue.
** nanogray (10^{-9} gray) per hour, per megabecquerel (10^6 becquerel)

4.5. Misadministration of Radiopharmaceutical

The administration of the wrong radiopharmaceutical or the wrong amount is a rare occurrence in the practice of nuclear medicine. The primary contributing factors appear to be errors associated with: (i) inadequate labelling and identification of radiopharmaceuticals; (ii) the lack of, or uncritical, processing of nuclear medicine requisitions, and (iii) incorrect patient identification.

After a misadministration of a diagnostic dose, prompt attention should be given to the care of the patient and to notification of appropriate individuals, depending on the existing local, regional or national rules or regulations. The anticipated absorbed dose should be calculated as quickly as possible. Efforts to reduce the potential absorbed dose, as given in Table 4, should be instituted immediately if feasible. After a misadministration of a therapeutic dose instead of a diagnostic dose, the procedures in Section 5.8 should be followed.

5. NUCLEAR MEDICINE IN THERAPY

Radiation therapy with unsealed sources is used for the treatment of both benign and malignant diseases. It is likely that the therapeutic use of radionuclides will increase in the future because of new developments in immunotherapy, antibody therapy, and in methods utilising unique metabolic characteristics of cancers.

Patients treated with radionuclides have been extensively studied for evidence of radiation-induced cancers or other detriment.

Patients with polycythemia vera treated with ^{32}P have been reported to be at increased risk of leukaemia, although a large part, if not all, of the increase in risk can be attributed to the course of the disease itself. Chemotherapy of polycythemia vera is accompanied by a significantly higher incidence of leukaemia than occurs following its treatment by ^{32}P.

In patients treated for hyperthyroidism with ^{131}I, compared with those treated by surgery, there was no evidence of an increase in leukaemia in those patients who received the radionuclide therapy. In thyroid cancer patients treated with ^{131}I, an excess in the incidence of and mortality from leukaemia, bladder and kidney cancers, and possibly of breast cancer was found.

Depending on the radionuclide and the activity administered, deterministic effects can follow radionuclide therapy. Examples include: (i) hypothyroidism after ^{131}I therapy for hyperthyroidism and (ii) bone marrow depression after ^{32}P therapy for leukaemia, malignant effusions, breast cancer and polycythemia.

5.1. Responsibilities of the Physician and the Medical Physicist

The responsibility for this form of radiotherapy rests with a physician who has proper training and sufficient knowledge of nuclear medicine therapy and of alternative therapeutic methods such as surgery, chemotherapy and hormonal therapy. The physician should be aware of the relative risks and benefits of all protocols and planned therapeutic methods.

Measurement of radionuclide activity, identification of radionuclides, and internal radiation dosimetry are the responsibilities of a medical physicist, who should also be concerned with radiation protection of the staff, of the immediate family, of other patients and of members of the public.

5.2. Treatment of Malignant Diseases

Treatment intended to cure a malignant disease, or to control its symptoms, may involve absorbed doses to normal tissues that approach or exceed the threshold levels for deterministic effects.

In deciding whether radionuclide therapy is the best treatment for a patient with cancer, consideration must be given to the risk of death or disability, or both, from the disease, as compared with the risk of injury from the irradiation. The treatment of malignant disease in children requires particularly careful evaluation of the potential benefits and risks. This special consideration is because of the sensitivity of a child's growing tissues to deterministic damage and the long potential life span in which tissue damage may be expressed.

To aid in treatment planning, it is often helpful to use a relatively small test activity to obtain information on the tissue distribution and on the effective half-life of the radiopharmaceutical in various organs and tissues of the individual patient.

5.3. Treatment of Benign Diseases

For benign diseases, the age of the patient is an important consideration in deciding whether radionuclide therapy is more appropriate than it is in the treatment of cancer.

5.4. Nuclear Medicine Therapy during Pregnancy

Because certain radiopharmaceuticals, including ^{131}I as iodide and ^{32}P as phosphate, can rapidly cross the placenta, the possibility of pregnancy should be very carefully considered before such radionuclides are given for therapy. As a rule, a pregnant woman should not be treated with a radioactive substance unless the therapy is required to save her life; in that event, the potential absorbed dose to the fetus should be estimated and consideration should be given to terminating the pregnancy.

5.5. Nuclear Medicine Therapy in Women of Reproductive Capacity

Pecautions that need to be taken to avoid treatment with radionuclides of women with undiagnosed pregnancy are specified in Section 4.3.1.

In addition, it is advisable that a woman not attempt to become pregnant until the activity remaining in the woman's body is such that it will not cause the absorbed dose in the conceptus to exceed 1 milligray.

5.6. Protection of the Family and Public after Nuclear Medicine Therapy

The physician responsible should provide necessary information to family members for their radiation protection. After administration of a therapeutic amount of radiopharmaceutical, the absorbed dose rate may be such that a relative remaining close to a patient for several days could receive an absorbed dose in the order of several tens of milligrays. Patients should not be released from the hospital until it can be expected that the incidental irradiation of members of the family will not give rise to absorbed doses exceeding 1 milligray. This limit does not apply to those who attend the patient at home.

A patient treated with gamma-emitting radionuclides should be advised not to hold children or otherwise be in intimate contact with family members for an appropriate time interval after having left the hospital. If the patient is a nursing mother, breast-feeding may have to be stopped after administration of a therapeutic amount of activity.

5.7. Incidental Irradiation of One Patient by Another

Patients given therapeutic amounts of gamma-emitting radiopharmaceuticals should preferably be accommodated in a separate room, which should not be accessible to patients untreated with radiation and properly shielded if necessary. Whenever practicable, toilets and similar facilities should be separate, and there should be frequent removal of radioactive waste from the ward.

5.8. Therapeutic Misadministration

Misadministration involves erroneous administration of an activity substantially higher (or lower) than required for a particular treatment. It includes also administration of a therapeutic amount to a patient not requiring treatment at all.

When a therapeutic misadministration (exceeding the required activity) is recognised, the nuclear medicine physician should immediately use all available means to minimise any adverse effects. Among these are: (i) the expeditious removal of orally administered radiopharmaceuticals by emesis, gastric lavage, laxatives or enemas; (ii) the accelerated excretion of intravenously administered radiopharmaceuticals by hydration, diuresis and chelation therapy when appropriate; (iii) the removal of urine by catheterisation from patients who cannot void spontaneously; (iv) the use, when appropriate, of blocking agents, such as KI or $KClO_4$, to diminish the absorbed dose to the thyroid gland, the salivary glands and the stomach.

After any therapeutic misadministration, it is essential that the patient and his family, or representative, be promptly notified. It may be necessary to take measures to ensure that family members, or others who may come into contact with the patient, are not unduly irradiated. The referring physician, hospital administration and other responsible authorities at local, regional or national level should also be notified, depending upon existing rules or regulations.

6. NUCLEAR MEDICINE PROCEDURES IN MEDICAL RESEARCH

6.1. With Direct Benefits to the Individual Irradiated

Nuclear medicine procedures forming part of a medical research program sometimes involve direct benefits for the irradiated individual and sometimes do not. When new and experimental methods of nuclear medicine are capable of benefiting the individuals on whom they are tested, the justification for the procedures can be judged in the same way as for other medical examinations. Nevertheless, because of the experimental character of the procedures, they should be subject to thorough review.

6.2. Without Direct Benefits to the Individual Irradiated

The decision to use nuclear medicine procedures on persons, for the purpose of those research and other studies in which no direct benefit to the persons irradiated is intended, should only be undertaken by specially qualified and trained research personnel and nuclear medicine physicians.

The estimated risks of the irradiation should be explained to those involved, who should be volunteers fully able to exercise their free will. The higher the dose the more rigorous should be the requirements on the conditions of securing volunteers and on their ability to understand the risk. Dose constraints should be considered when the medical research procedures are not intended to be of direct value to the irradiated individual.

Such irradiation should only be given with the consent of the authorities in charge of the institution where the irradiation is to take place, as advised by an appropriate expert body and subject to local and national regulations.

The Commission has produced a report on *Radiological Protection in Biomedical Research* (ICRP, 1992) which gives guidance on this topic.

7. RADIOPHARMACEUTICALS: PREPARATION AND QUALITY CONTROL

7.1. Radiopharmaceuticals and Radiopharmacy

The purpose of quality control in radiopharmacy is to ensure that the radiopharmaceuticals are effective, toxicologically safe, and contain known ingredients of known quality in known proportions and predetermined purity. The highest standards of production of radiopharmaceuticals are required.

7.2. Sterility

Pharmaceuticals to be labelled with short-lived radionuclides should be tested for compliance with respective standards prior to labelling on the premises. Should any doubt arise as to the full sterility of the final product, micropore filters for additional sterilisation should be employed prior to injection.

7.3. Quality Control

Quality control of the preparation of radiopharmaceuticals is essential to the proper use of the radiopharmaceuticals and should be supervised by properly trained personnel. Each individual amount of activity to be administered should be assayed in a calibrated activity meter prior to the administration of the material in addition to the quality control procedures identified below for specific classes of radiopharmaceuticals.

7.3.1. *Ready-for-use radiopharmaceuticals*

Upon receipt of the product it is essential to compare carefully the details of the written specification sheet, the label, and the package documentation. The container should be carefully examined, using good radiation protection measures, to be sure that it is not leaking.

7.3.2. *Radionuclide generators*

For injection, it is necessary to prepare the eluate under sterile conditions. The elution yield from the generator should be calculated and compared with a measurement as an indication of whether or not the generator is working properly.

7.3.3. *Radionuclide kits*

The necessity for radiochemical purity testing of kit-prepared compounds on the type of radiopharmaceutical and the method of manufacture. The manufacturer's instructions for constituting the radiopharmaceutical should be strictly adhered to by qualified personnel. When the labelling is incomplete, a regular control, before administration of the radiopharmaceutical to patients, should be introduced and faulty preparations discarded. The manufacturer should be notified immediately of the situation. Kits should not be used beyond the expiry date.

7.3.4. *Facility-produced radiopharmaceuticals*

With facility-produced radiopharmaceuticals the facility itself is responsible for all aspects of product quality.

7.3.5. *Autologous labelled radiopharmaceuticals*

In addition to the recommendations for facility-produced radiopharmaceuticals, factors that must be considered include the assurance of the separation and viability of labelled cells, maintenance of sterility throughout the labelling process and the minimisation of the risk of infecting personnel.

8. NUCLEAR MEDICINE FACILITIES AND QUALITY CONTROL OF INSTRUMENTATION

8.1. Facilities

The primary objective of facility design is to enable radiopharmaceuticals to be prepared and administered in a safe manner. For this purpose, it is necessary to facilitate the control and removal of radioactive contamination, and, in particular, to minimise interference with radionuclide measuring and imaging devices by extraneous contamination from radiation sources, including patients themselves. Increases in the ambient radiation levels should be minimised by careful attention to the storage and movement of radioactive materials, including the position and movement of patients receiving such materials for therapy or diagnosis.

Special localised ventilating systems for removing radioactive gases and other volatile radiopharmaceuticals from imaging rooms may be required to prevent cross-contamination that might affect other detection systems.

There should be a special room for administering radiopharmaceuticals to patients. The floor of this room should be easy to decontaminate.

Radiation emitted from patients containing large quantities of radioactive materials can interfere with measurements being made on other patients. Areas for treating patients should therefore be separate from the rest of the laboratory, and located so that the treated patients need not pass through or close to the rest of the laboratory after having received the activity.

There should be a special room for keeping any radioactive waste that needs to be stored.

8.2. Quality Control for Nuclear Medicine Equipment

The remainder of this section deals with quality control for nuclear medicine equipment, and the quantitative tests that need to be carried out for each class of instrument to ensure its initial and continued efficiency.

8.2.1. *Acceptance testing*

Results from acceptance tests may be used for comprehensive assessments of future performance. Reference tests should be repeated after any repair of a major failure, or when an instrument is moved to a new site.

8.2.2. *Routine testing and maintenance*

Routine quantitative tests should be carried out regularly, to ensure that each instrument continues to perform as required, or to determine the rate and extent of the deterioration of performance with time. All personnel should be on the alert for instrument malfunction and the presence of artifacts.

8.2.3. *Instrument records*

A data-recording log book is essential for monitoring the quality control and servicing requirements of each instrument.

8.2.4. *Radiation source requirements for various tests*

It is necessary to have calibrated gamma-emitting radiation sources of appropriate energy, activity and shape. The calibration should be traceable to appropriate national or other standards.

8.2.5. *Activity meters*

Quality control of these instruments must include consideration of background radiation, shield leakage, linearity of response, and the required accuracy of measurements and precision of the measurements for radionuclides and geometries of equipment (vials and syringes) in current use.

8.2.6. *Well counters*

Quality control of counting systems for gamma radiation measurements *in vitro* includes tests for the performance of the pulse height analyser, scaler or ratemeter. It is necessary to check the linearity of energy response, energy resolution, sensitivity for radionuclides in current use, counting precision, background and proportionality between count rate and measured activity.

8.2.7. *Probe systems*

In addition to the tests indicated above, tests on output devices, collimators, and multiprobe systems seen to be undertaken on each individual probe and its associated electronic channel.

8.2.8. *Quality control of photographic processes*

The photographic process is critical in assuring optimal image quality in nuclear medicine. It is necessary to check daily the settings of the formatter and video hard-copy (multi-format) camera. The film processor should also be checked daily.

8.2.9. *Gamma cameras*

Quality control, to determine the rate and extent of fluctuation, and deterioration with time, of various performance parameters, is important for gamma cameras because of their electronic complexity. The performance tests should include calibration of the analyser energy peak and window controls, determination of the energy resolution, spatial resolution, uniformity, sensitivity, spatial linearity, count-rate performance and shield leakage. A quality assurance program should include routine mechanical and electrical inspections, and the display units should be checked.

8.2.10. *Single photon emission computed tomographic (SPECT) systems*

SPECT scintigraphy is significantly more demanding than planar scintigraphy. The four factors that can have the most significantly adverse impact on the diagnostic quality of the images are camera non-uniformity, misalignment of the geometric axis of rotation with the electronic axes of the camera head, mechanical instability and electronic instability.

SPECT acquisition data must be corrected for camera non-uniformity, and for any misalignment of the camera head with respect to the central axis of rotation. The use of larger quantities of administered activity, compared with those used in conventional imaging, may be justified, in order to reduce the noise level in SPECT imaging. Factors that affect the spatial resolution in SPECT images include the type of collimator, the amount of filtering and the radius of rotation. This last factor must be minimised to achieve the best resolution.

Mechanical checks should be performed as part of acceptance. Collimator shift and deformation should be noted. Vibration should be monitored to ensure that camera-head vibration has stopped before SPECT data acquisition has started.

It is necessary to establish at least weekly the following conditions for proper operation of a SPECT system:

— Angular reliability and detector tilt.

— Electronic centre of rotation: the centre of rotation should be established to within $\frac{1}{4}$ of a pixel (usually) or 1.5 mm.

— Uniformity: the uniformity correction matrix should be performed carefully for each collimator and for each radionuclide to be used with that collimator. Approximately 3×10^7 counts should be used in acquiring the uniformity correction image.

8.2.11. *Positron-emission tomography*

The energy of annihilation gamma rays (0.511 MeV) is higher than that of the radiation involved in most nuclear medicine procedures, and so requires more shielding of the imaging room and other space, for protection of the patient.

8.2.12. *Data-processing systems*

The equipment should be checked against the manufacturer's specifications before acceptance.

Reference Publications of the International Commission on Radiological Protection (ICRP)

ICRP (1977). *Recommendations of the ICRP*, ICRP Publication 26. *Annals of the ICRP* **1**(3).

ICRP (1987a). *Protection of the Patient in Nuclear Medicine*, ICRP Publication 52. *Annals of the ICRP* **17**(4).

ICRP (1987b). *Radiation Dose to Patients from Radiopharmaceuticals*, ICRP Publication 53. *Annals of the ICRP* **19**(1–4).

ICRP (1991). *1990 Recommendations of the International Commission on Radiological Protection*, ICRP Publication 60. *Annals of the ICRP* **21**(1–3).

ICRP (1992). *Radiological Protection in Biomedical Research* and *Addendum 1 to Publication 53—Radiation Dose to Patients from Radiopharmaceuticals*. ICRP Publication 62. *Annals of the ICRP* **22**(3).